Thomas W. Wieting
Reed College, 1970.

Functional Analysis

Functional Analysis

Proceedings of a Symposium
held at Monterey, California
October 1969

Edited by
Carroll O. Wilde

Naval Postgraduate School
Monterey, California

ACADEMIC PRESS
NEW YORK AND LONDON
1970

ACADEMIC PRESS, INC.
111 Fifth Avenue, New York, New York 10003

United Kingdom Edition published by
ACADEMIC PRESS, INC. (LONDON) LTD.
Berkeley Square House, London W1X 6BA

LIBRARY OF CONGRESS CATALOG CARD NUMBER: 78-117085
AMS 1968 SUBJECT CLASSIFICATION 4600, 4700

PRINTED IN THE UNITED STATES OF AMERICA

CONTENTS

PREFACE

From its inception the field of functional analysis has been useful in solving problems in applied mathematics. Indeed, many of the early studies in functional analysis centered around problems in integral equations. The subject has been studied extensively in recent years, and the study has resulted in well-organized, carefully formulated and useful mathematical theories, as well as the establishment of connections with other areas of mathematics.

Many investigators in important areas of applied mathematics today feel the need for the ideas and knowledge already accumulated within functional analysis, and for new concepts currently being developed. Researchers in various areas are often hampered by a lack of knowledge of the concepts and precise formulations of functional analysis, and with increasing regularity they find the new and significant results in their own fields written in the language and within the framework of functional analysis.

The purpose of this symposium was twofold. One aim was to promote further development of the field of functional analysis itself, but the primary goal was to bring to the attention of investigators in various areas of mathematics some of the most recent results in functional analysis and closely related topics. Thus several of the papers in these proceedings are devoted to functional analysis *per se*, while others are devoted to closely related topics, such as real analysis and ergodic theory, and explore connections with functional analysis.

I take this opportunity to thank the Office of Naval Research for its support of the symposium and the Naval Postgraduate School for support and cooperation in the use of its facilities. I should also like to thank the members of the organizing committee, R. V. Chacon, S. Kakutani, G. K. Kalisch and G.-C. Rota, for their work done in planning the conference.

The camera-ready copy of the manuscript was prepared on an IBM Selectric Composer, and IBM Corporation was most generous and cooperative in making the machine available. Mostly I wish to thank Roxanne Maday for typing the copy -- she did an excellent job without the benefit of prior technical experience or prior experience on the Composer.

The proofreading was shared by several members of the Mathematics Department of the Naval Postgraduate School, D. L. Davis, R. W. Preisendorfer, D. W. Robinson, A. B. Schwarzkopf, A. M. Shorb, M. D. Weir, and the editor.

SEVERAL VARIABLE SPECTRAL THEORY[1]

Joseph L. Taylor[2] – *University of Utah*

The purpose of this article is to outline some recent results in the spectral theory of n-tuples of mutually commuting operators on a Banach space. We begin by outlining the problem that interests us and reviewing some of its classical solutions.

Let X be a Banach space and L(X) the Banach algebra of all bounded linear operators on X. If A is a topological algebra over the complex field $\mathbb{C}$, then an *action* of A on X is a continuous homomorphism of A into L(X) which carries the identity to the identity operator. In other words, an action of A on X is a left A-module structure on X such that the map $(a,x) \to ax : A \times X \to X$ is continuous. We submit that spectral theory is primarily concerned with the following problem:

FUNCTIONAL CALCULUS PROBLEM. Given topological algebras $A \subset B$, with A containing the identity of B, when does an action of A on a Banach space X extend to an action of B on X?

In the most common situation, A is the algebra P of complex polynomials in a single variable. We consider P an algebra of functions on $\mathbb{C}$ and give it the compact-open topology. An action of P on X is determined by specifying the operator b which is the image in L(X) of the function z. The action of P on X is then given by the homomorphism $p \to p(b) : P \to L(X)$.

If U is a domain in $\mathbb{C}$ and $\mathfrak{A}(U)$ is the algebra of functions analytic in U with the compact-open topology, then P is a subalgebra of $\mathfrak{A}(U)$. For this pair of algebras the above problem has a very well known solution: an action $p \to p(b)$ of P on X extends to an action of $\mathfrak{A}(U)$ on X if and only if U contains the spectrum Sp(b) of the operator b. Furthermore, the extension is unique and is given by $f \to f(b)$, where

$$f(b) = \frac{1}{2\pi i} \int_{\partial D} f(z)(z - b)^{-1} dz$$

for any domain D with "nice" boundary such that $Sp(b) \subset D \subset \bar{D} \subset U$.

---

[1] Research partially supported by the Air Force Office of Scientific Research, Office of Aerospace Research, United States Air Force, under AFOSR grant number 1313-67.

[2] The author was a fellow of the Alfred P. Sloan Foundation while portions of this research were being conducted.

If X happens to be a Hilbert space and b a normal operator on X, then the action of P on X, determined by b, extends all the way to an action of the algebra of bounded Borel functions on Sp(b). In fact, integration with respect to the spectral measure for b gives such an extension (cf. [3], Chapter X).

The above results form the foundation of operator theory. However, they solve the functional calculus problem for only the most trivial pairs of algebras A and B. The problem is equally valid and interesting if A is the algebra of polynomials in several variables rather than one, or if A is an arbitrary commutative algebra, or even if A is a noncommutative algebra. In this article we shall be concerned with the case where A is the algebra P_n of polynomials in n variables, and B the algebra $\mathfrak{A}(U)$ of functions analytic on a domain $U \subset \mathbb{C}^n$.

An action of P_n on X is determined by an n-tuple $a = (a_1, \ldots, a_n)$ of mutually commuting operators on X, where the action of the ith coordinate function z_i is determined by the operator a_i. If we expect results analogous to those for a single variable, we need a notion of spectrum for such an n-tuple. A well known approach to this problem is supplied by commutative Banach algebra theory. If A is a closed commutative subalgebra of L(X) with $a_1, \ldots, a_n \in A$, then the tuple $a = (a_1, \ldots, a_n)$ is said to be nonsingular (with respect to A) if the equation

$$a_1 b_1 + \ldots + a_n b_n = I \tag{1}$$

has a solution for $b_1, \ldots, b_n \in A$. The spectrum $Sp_A(a)$ of a with respect to A is then the set of all $z = (z_1, \ldots, z_n) \in C^n$ such that the tuple $z - a = (z_1 - a_1, \ldots, z_n - a_n)$ is singular (we write $z_i - a_i$ for $z_i I - a_i$). The set $Sp_A(a)$ is compact and non-empty, and the Shilov-Arens-Calderon Theorem (cf. [1], [2], [7]) states that the action of P_n on X extends to an action $f \to f(a)$ of $\mathfrak{A}(U)$ on X whenever U is a domain in $\mathbb{C}^n$ containing $Sp_A(a)$. In fact, each $f(a)$ is an element of A in this case.

Note that the above procedure has some obvious defects. The set $Sp_A(a)$ depends strongly on the choice of the algebra A, whereas it should depend only on $a_1, \ldots, a_n$. Also, to say that $z \notin Sp_A(a)$ involves hypothesizing the existence of another tuple $(b_1, \ldots, b_n)$ of operators. The functional calculus is supposed to allow one to conclude that operators with certain properties exist. Its value is diluted if we have to hypothesize the existence of enormous numbers of other operators. If b is a single operator and $z \in C$, then $z \notin Sp(b)$ means that the operator $z - b$ is one-to-one and onto; compare the simplicity of this condition with the problem posed by the solvability of equation (1).

We propose a notion of spectrum for an n-tuple a which avoids the difficulties mentioned above. This notion was introduced in [8]. In [9] we proved the corresponding analytic functional calculus theorem. We shall outline the results of these papers as well as some more recent insights in the following four sections. However, before proceeding with this outline, we shall introduce our new notion of spectrum by describing it in the situation where $a = (a_1, a_2)$, a pair of commuting operators.

Consider the two systems of equations

$$(2) \quad \begin{cases} a_1 x = u_1 \\ a_2 x = u_2 \end{cases} \qquad \text{and} \qquad (3) \qquad a_1 y_2 - a_2 y_1 = v,$$

where u_1, u_2 and v are given elements of X. Note that $a_1 a_2 = a_2 a_1$ implies that a necessary condition for (2) to have a solution is that $a_1 u_2 - a_2 u_1 = 0$. Also, a solution to (3) cannot be unique, since each pair of the form $y_1 = a_1 x$, $y_2 = a_2 x$ satisfies the homogeneous equation. Hence, the best that one can hope for in the way of existence and uniqueness for the above systems is the following:

(2) has a unique solution whenever $a_1 u_2 - a_2 u_1 = 0$; and

(3) has a solution for arbitrary u, which is unique modulo pairs $(y_1, y_2) = (a_1 x, a_2 x)$.

It seems reasonable to call the pair (a_1, a_2) *nonsingular* when this situation attains.

We define maps $a^0 : X \to X \oplus X$ and $a^1 : X \oplus X \to X$ as follows:

$$a^0 x = (a_1 x, a_2 x) \text{ and } a^1(y_1, y_2) = a_1 y_2 - a_2 y_1.$$

The existence and uniqueness criteria of the above paragraph can then be restated as $\operatorname{Ker} a^0 = 0$, $\operatorname{Ker} a^1 = \operatorname{Im} a^0$, and $\operatorname{Im} a^1 = X$; that is, the sequence

$$(4) \qquad 0 \to X \overset{a^0}{\to} X \oplus X \overset{a^1}{\to} X \to 0$$

is exact. Note that $a_1 a_2 = a_2 a_1$ implies that (4) is always a complex (the kernel of each map contains the image of the preceding map).

If we agree that a commuting pair (a_1, a_2) is nonsingular if (4) is exact, and define the ***spectrum*** of (a_1, a_2) to be the set of $(z_1, z_2) \in \mathbb{C}^2$ such that $(z_1 - a_1, z_2 - a_2)$ is singular, then we have a notion of spectrum which does not involve hypothesizing the existence of additional operators. Of course, it remains to extend this definition to the case of n-tuples for $n > 2$, show that it has certain basic properties, and develop a corresponding analytic functional calculus. We outline how this is done in the remaining sections.

§1. DEFINITION AND ELEMENTARY PROPERTIES OF THE SPECTRUM.

Throughout this section X is a Banach space and $a = (a_1, \ldots, a_n)$ a commuting n-tuple of bounded linear operators on X. We refer the reader to [6], Chapter XVI, and [5], Chapter 2, for background information on exterior algebra and cochain complexes.

We choose indeterminates $e_1, \ldots, e_n$ and let $\Lambda[e_1, \ldots, e_n]$ denote the exterior algebra (over $\mathbb{C}$) generated by $e_1, \ldots, e_n$. The subspace of $\Lambda[e_1, \ldots, e_n]$ consisting of elements of degree p will be denoted $\Lambda^p[e_1, \ldots, e_n]$. A typical element of $\Lambda^p[e_1, \ldots, e_n]$ has the form

$$\sum_{1 \leqslant j_1, \ldots, j_p \leqslant n} \lambda_{j_1 \ldots j_p} e_{j_1} \wedge \ldots \wedge e_{j_p} \quad (\lambda_{j_1 \ldots j_p} \in \mathbb{C})$$

for $p > 0$, while $\Lambda^0[e_1, \ldots, e_n] = \mathbb{C}$ and $\Lambda^p[e_1, \ldots, e_n] = 0$ for $p < 0$.

We next define a cochain complex

$$(1.1) \qquad 0 \to F^0(X, a) \overset{a^0}{\to} F^1(X, a) \overset{a^1}{\to} F^2(X, a) \overset{a^2}{\to} \cdots .$$

The set $F^p(X, a)$ of p-cochains is defined to be $X \otimes \Lambda^p[e_1, \ldots, e_n]$, where a typical element ψ will be written in the form

$$\psi = \Sigma\, x_{j_1 \ldots j_p} e_{j_1} \wedge \ldots \wedge e_{j_p} \quad (x_{j_1 \ldots j_p} \in X)$$

for $p > 0$, while $F^0(X, a) = X$. The coboundary operator $a^p : F^p(X, a) \to F^{p+1}(X, a)$ is defined by

$$a^p(\Sigma\, x_{j_1 \ldots j_p} e_{j_1} \wedge \ldots \wedge e_{j_p}) = \sum_{i=1}^{n} \sum_{j_1 \ldots j_p} (a_i x_{j_1 \ldots j_p}) e_i \wedge e_{j_1} \wedge \ldots \wedge e_{j_p}$$

for $p > 0$, while $a^0 x = (a_1 x)e_1 + \ldots + (a_n x)e_n$ (note the analogy with differential forms). If we write the tuple $a = (a_1, \ldots, a_n)$ in the form $a = a_1 e_1 + \ldots + a_n e_n$, then formally $a^p \psi = a \wedge \psi$ for $\psi \in F^p(X, a)$.

A simple computation using the commutativity of $a_1, \ldots, a_n$ shows that (1.1) is indeed a cochain complex ($a^p \circ a^{p-1} = 0$ for each p).

1.1. DEFINITION. We denote the complex (1.1) by $F(X, a)$. Its cohomology, $H^p(X, a) = \{H^p(X, a)\}_p$, is defined by $H^p(X, a) = \operatorname{Ker} a^p / \operatorname{Im} a^{p-1}$. We shall say that a is *nonsingular* provided $H(X, a) = 0$, i.e., provided the complex $F(X, a)$ is exact. The *spectrum*, $\operatorname{Sp}(a, X)$, of a is the set of $z \in \mathbb{C}^n$ for which the tuple $z - a = (z_1 - a_1, \ldots, z_n - a_n)$ is singular.

Therefore, $z \in \operatorname{Sp}(a, X)$ if and only if for some p the cohomology group $H^p(X, z - a_n)$ is singular.

Note that if $a = (a_1)$ is a 1-tuple, then the complex $F(X, z - a)$ is just $0 \to X \xrightarrow{z - a_1} X \to 0$. In this case, $H^0(X, z - a) = \operatorname{Ker}(z - a_1)$ is the set of eigenvectors for a_1 and $H^1(X, z - a) = X/\operatorname{Im}(z - a_1) = \operatorname{Coker}(z - a_1)$. If $a = (a_1, a_2)$ is a pair, then the complex $F(X, a)$ is the complex (4) discussed in the introduction.

For an n-tuple $a = (a_1, \ldots, a_n)$ it is always the case that $H^0(X, z - a) = \operatorname{Ker}(z - a)^0 = \{x \in X : (z_1 - a_1)x = 0,\ i = 1, \ldots, n\}$. Hence, $H^0(X, z - a)$ is the set of common eigenvectors for $a_1, \ldots, a_n$ with eigenvalues $z_1, \ldots, z_n$. One might think of elements of $H^1(X, z - a)$ as common eigenvectors once removed, with similar statements for $H^p(X, z - a)$ with $p > 1$. This viewpoint is justified by the following proposition:

1.2. PROPOSITION ([8], Lemma 1.1). Let $(a)'$ denote the algebra of all operators on X which commute with each a_i. The action of $(a)'$ on X then induces a natural action of $(a)'$ on $H^p(X, z - a)$ for each $z \in C$ and each p. Under this action, the operators $a_1, \ldots, a_n$ act as multiplication by the scalars $z_1, \ldots, z_n$ respectively.

The above action of $(a)'$ is defined as follows: each $F^p(X, z - a)$ is, in a natural way, a direct sum of $\binom{n}{p}$ copies of X; hence $F^p(X, z - a)$ is an $(a)'$-module. The maps $(z - a)^p : F^p(X, z - a) \to F^p(X, z - a)$ are $(a)'$-module homomorphisms, since $a_1, \ldots, a_n$ are in the center of $(a)'$. It therefore follows that $H^p(X, z - a) = \mathrm{Ker}(z - a)^p / \mathrm{Im}(z - a)^{p-1}$ is an $(a)'$-module, and a simple computation shows that if $\psi \in \mathrm{Ker}(z - a)^p$, then $(z_i - a_i)\psi \in \mathrm{Im}(z - a)^{p-1}$ for each i. Hence, a_i acts as z_i on $H^p(X, z - a)$.

We return briefly to the classical notion of joint spectrum. If A is any subalgebra of L(X) with $a_1, \ldots, a_n$ in its center, we say that $z \in \mathrm{Sp}_A(a)$ if the equation

$$(z_1 - a_1)b_1 + \ldots + (z_n - a_n)b_n = I \tag{1.2}$$

fails to have a solution for $b_1, \ldots, b_n \in A$. Note that if A is norm closed, then $\mathrm{Sp}_A(a)$ is a subset of $\{z \in \mathbb{C}: |z_1| \leqslant r_i,\ i = 1, \ldots, n\}$, where r_i is the spectral radius of a_i. Since $(a)'$ is the largest subalgebra of L(X) containing $a_1, \ldots, a_n$ in its center, the set $\mathrm{Sp}_A(a)$ will be minimal if we choose $A = (a)'$. The following proposition is an immediate consequence of Proposition 1.2:

1.3. PROPOSITION. If A is any subalgebra of L(X) with $a_1, \ldots, a_n \in$ center (A), then $\mathrm{Sp}(a, X) \subset \mathrm{Sp}_A(a)$.

An example in §4 of [8] shows that the above containment can be proper even if A is chosen to be $(a)'$.

Since $F^p(X, a)$ is a direct sum of $\binom{n}{p}$ copies of X, it is a Banach space. The maps $(z - a)^p$ are bounded linear maps which depend continuously - in fact, analytically - on the parameter z. These facts together with the open mapping theorem can be used to prove the following:

1.4. PROPOSITION ([8], Theorem 3.1). The set of $z \in \mathbb{C}^n$ for which the complex $F(X, z - a)$ is exact is an open set. Hence, $\mathrm{Sp}(a, X)$ is compact.

The next proposition specifies what happens to $\mathrm{Sp}(a, X)$ if the tuple $a = (a_1, \ldots, a_n)$ is permuted or has some of its entries deleted.

1.5. PROPOSITION ([8], Theorem 3.2). Suppose $m < n$ and $(k_1, \ldots, k_m)$ is an m-tuple of distinct integers between 1 and n. If $k^*a = (a_{k_1}, \ldots, a_{k_m})$ and $k^*z = (z_{k_1}, \ldots, z_{k_m}) \in \mathbb{C}^m$ for $z = (z_1, \ldots, z_n) \in \mathbb{C}^n$, then $\mathrm{Sp}(k^*a, X) = k^*\mathrm{Sp}(a, X)$.

This is a key result. It implies, among other things, that $\mathrm{Sp}(a, X)$ is always nonempty. In fact, if $\mathrm{Sp}(a, X)$ were empty, then Proposition 1.5 would imply that the spectrum of each of the single operators $a_1, \ldots, a_n$ was empty, which is impossible.

Suppose now that $X_0 \subset X$ is a closed subspace which is invariant under each a_i. We may then consider $a = (a_1, \ldots, a_n)$ to be a tuple of operators on X_0 and X/X_0 as well as on X. The short exact sequence $0 \to X_0 \to X \to X/X_0 \to 0$ induces a short exact sequence $0 \to F(X_0, a) \to F(X, a) \to F(X/X_0, a) \to 0$ of cochain complexes, which induces a corresponding long exact sequence of cohomology (cf. [5], Chapter 2). An inspection of this sequence proves the following:

1.6. PROPOSITION ([8]. Theorem 3.4). If a is nonsingular on any two of X_0, X, or X/X_0, then it is nonsingular on the third as well. Hence the union of any two of $Sp(a, X_0)$, $Sp(a, X)$, or $Sp(a, X/X_0)$ contains the third.

We consider the above proposition to be one of the main virtues which distinguishes our notion of spectrum from others. It is a valuable tool in computing the spectrum of specific n-tuples.

§2. THE ANALYTIC FUNCTIONAL CALCULUS. We mentioned in the introduction that our main interest in spectral theory is the problem of extending an action of an algebra on a Banach space to an action of a larger algebra. In this section we outline how our notion of spectrum relates to this problem in the case of algebras of analytic functions. Proofs of the theorems can be found in [9].

If U is any domain in $\mathbb{C}^n$, let $\mathfrak{A}(U)$ denote the algebra of functions analytic on U. We give $\mathfrak{A}(U)$ the topology of uniform convergence on compact sets. An action of $\mathfrak{A}(U)$ on a Banach space X is a continuous homomorphism of $\mathfrak{A}(U)$ into $L(X)$ which carries the identity to the identity.

Recall that an action of the polynomial algebra P_n in n-variables is uniquely determined by the commuting n-tuple of operators $(a_1, \ldots, a_n)$, where a_i is the image of the coordinate function z_i.

Our main theorem generalizes the Shilov-Arens-Calderon Theorem for commutative Banach algebras (cf. [1], [2], [7]). It is proved in [9] using an adaptation of the Cauchy-Weil integral.

2.1. THEOREM ([9], Theorem 4.3). If $a = (a_1, \ldots, a_n)$ is a commuting n-tuple of operators on X, then the action $p \to p(a)$ of P_n on X extends to an action $f \to f(a)$ of $\mathfrak{A}(U)$ for each domain $U \supset Sp(a, X)$. Furthermore, $f(a) \in (a)''$ for each $f \in \mathfrak{A}(U)$, where $(a)''$ is the algebra of operators in $L(X)$ which commute with each element of $(a)'$.

If $a_1, \ldots, a_n$ are elements of commutative Banach algebra A, then one has an alternate characterization of $Sp_A(a)$ as $\{(h(a_1), \ldots, h(a_n)): h \in \Delta\}$, where Δ is the set of complex homomorphisms of A (cf. [3]). This fact makes the Banach algebra version of the next proposition fairly easy to prove. However, in our situation the proof is surprisingly difficult.

2.2. PROPOSITION ([9], Theorem 4.8). Let U be a domain containing $Sp(a, X)$ and $f \to f(a)$: $\mathfrak{A}(U) \to L(X)$ the action guaranteed by Theorem 2.1. If $f = (f_1, \ldots, f_m)$ is an m-tuple of elements of $\mathfrak{A}(U)$ and $f(a) = f_1(a), \ldots, f_m(a))$, then $Sp(f(a), X) = \{(f_1(z), \ldots, f_n(z)): z \in Sp(a, X)\} = f(Sp(a, X))$.

The Shilov idempotent theorem [7] is a standard application of the Shilov-Arens-Calderon Theorem. We obtain a strengthened version of this result by a combining (2.1) and (2.2).

2.3. PROPOSITION ([9]. Theorem 4.9). If $Sp(\alpha, X) = K_1 \cup K_2$, where K_1 and K_2 are disjoint compact sets, then $X = X_1 \oplus X_2$, where X_1 and X_2 are closed subspaces invariant under each element of $(\alpha)'$ with $Sp(\alpha, X_1) = K_1, Sp(\alpha, X_2) = K_2$.

The spaces X_1 and X_2 above are obtained as the kernel and image respectively of the projection operator $f(\alpha)$, where f is a function analytic in a neighborhood of $Sp(\alpha, X)$ which is zero on K_1 and one on K_2.

Note that our definition of $Sp(\alpha, X)$ avoids hypothesizing the existence of a solution to the equation

$$(2.1) \qquad (z_1 - a_1)b_1 + \ldots + (z_n - a_n)b_n = I$$

when $z = (z_1, \ldots, z_n) \notin Sp(\alpha, X)$. However, using the analytic functional calculus, we we can prove that this equation does have a solution for certain points $z \in Sp(\alpha, X)$.

2.4. PROPOSITION ([9], Theorem 5.4). If $z \notin Sp(\alpha, X)$ has the property that the equation

$$(2.2) \qquad (z_1 - w_1)f_1(w) + \ldots + (z_n - w_n)f_n(w) = 1$$

has a solution for functions $f_1, \ldots, f_n$ analytic in a neighborhood of $Sp(\alpha, X)$, then equation (2.2) has a solution for $b_1, \ldots, b_n \in (\alpha)''$. In particular, if (2.2) has a solution for each $z \notin Sp(\alpha, X)$, then $Sp(\alpha, X) = Sp_{(\alpha)''}(\alpha)$.

Note that if U is a domain of holomorphy containing $Sp(\alpha, X)$, then (2.2) has a solution for $f_1, \ldots, f_n \in \mathfrak{A}(U)$ provided $z \notin U$ (cf. [4]).

§3. THE SPECTRUM OF AN ALGEBRA ACTION. In this section A will be a commutative topological algebra with identity over the complex numbers. We denote the set of continuous homomorphisms of A onto $\mathbb{C}$ by Δ. We set $a(h) = h(a)$ for $a \in A$, $h \in \Delta$ and give Δ the weak topology generated by $\{\hat{a}: a \in A\}$.

A Banach A-module is a Banach space X together with a continuous action $(a, x) \to ax$ of A on X. If $\alpha = (a_1, \ldots, a_n)$ is a tuple of elements of A, then $Sp(\alpha, X)$ will denote the spectrum of α considered as a tuple of operators on X.

If $\alpha = (a_1, \ldots, a_n)$ and $\beta = (b_1, \ldots, b_m)$ are tuples of elements of A, then we write $\beta \leq \alpha$ provided there is an m-tuple $(k_1, \ldots, k_m)$ of distinct integers between 1 and n such that $b_i = a_{k_i}$ for $i = 1, \ldots, m$. In this situation, according to Proposition 1.5, the map $k^*: \mathbb{C}^n \to \mathbb{C}^m (k^*(z_1, \ldots, z_n) = (z_{k_1}, \ldots, z_{k_m})$ maps $Sp(\alpha, X)$ onto $Sp(\beta, X)$. We denote this map by $\pi_{\alpha\beta}$. Note that $(\{Sp(\alpha, X)\}\alpha, \{\pi_{\alpha\beta}\}_{\beta \leq \alpha})$ forms an inverse limit system of compact sets with onto maps.

Now each n-tuple $a = (a_1, \ldots, a_n)$ determines a continuous map $\hat{a} = (\hat{a}_1, \ldots, \hat{a}_n): \Delta \to \mathbb{C}^n$. It follows from Proposition 1.3 that the image of a contains $Sp(a, X)$. We put $\Delta(A, X) = \bigcap_a \hat{a}^{-1}(Sp(a, X))$. Then the following is not difficult to prove (cf. [8], §3):

3.1. PROPOSITION. The set $\Delta(A, X)$ is compact and nonempty in Δ. Each $a: \Delta \to \mathbb{C}^n$ maps $\Delta(A, X)$ onto $Sp(a, X)$. The set of maps $\{\hat{a}: \Delta(A, X) \to Sp(a, X)\}_a$ induces a homomorphism between $\Delta(A, X)$ and $\varprojlim (\{Sp(a, X)\}_a, \{\pi_{a\beta}\}_{\beta \leqslant a})$.

Thus, each Banach A-module X determines a compact subset $\Delta(A, X)$ of Δ such that $\hat{a}$ maps $\Delta(A, X)$ onto $Sp(a, X)$ for each tuple a. It follows from Proposition 1.6 that the correspondence $X \to \Delta(A, X)$ has the following property:

3.2. PROPOSITION ([8], Theorem 3.4). If $0 \to X_1 \to X_2 \to X_3 \to 0$ is a short exact sequence of Banach A-modules, then the union of any two of $\Delta(A, X_1)$, $\Delta(A, X_2)$ or $\Delta(A, X_3)$ contains the third.

Proposition 3.2 implies, in particular, that if $\Delta(A, X_1)$ and $\Delta(A, X_3)$ are disjoint, then $\Delta(A, X_2) = \Delta(A, X_1) \cup \Delta(A, X_3)$. If we choose an n-tuple $a = (a_1, \ldots, a_n)$ such that $\hat{a}$ separates $\Delta(A, X_1)$ and $\Delta(A, X_3)$ and apply Propositions 3.1 and 3.2, we obtain:

3.3. PROPOSITION. Suppose that X_1 and X_3 are Banach A-modules such that $\Delta(A, X_1) \cap \Delta(A, X_3) = 0$; then any short exact sequence $0 \to X_1 \xrightarrow{\mu} X_2 \xrightarrow{\nu} X_3 \to 0$ of Banach A-modules splits; that is, ν has a right inverse which is a bounded linear map as well as an A-module homomorphism.

§4. CONNECTIONS WITH HOMOLOGICAL ALGEBRA. There is an interesting relationship between the notion of spectrum we have defined here and certain ideas in homological algebra. This relationship is interesting because it suggests a way that spectral theory might be extended to noncommutative situations.

Let A be an algebra with identity (over the complex field, although this is an unnecessary restriction) and let M be an A-bimodule. That is, M is a linear space together with bilinear maps $(a, m) \to am : A \times M \to M$ and $(m, a) \to ma: M \times A \to M$ satisfying $a(mb) = (am)b$, $a(bm) = (ab)m$, $(ma)b = m(ab)$, and $1m = m = m1$. We shall describe the Hochschild cohomology of A with coefficients in M (cf. [5]).

4.1. DEFINITION (Hochschild). With A and M as above, let $C^p(A, M)$ denote the A-bimodule consisting of all p-multilinear maps $f: \overset{p}{\times} A \to M$ for $p > 0$. Let $C^0(A, M) = M$. We define a map $\delta^p: C^p(A, M) \to C^{p+1}(A, M)$ by

$$(4.1) \quad \delta^p f(a_1, \ldots, a_{p+1}) = a_1 f(a_2, \ldots, a_{p+1}) + (-1)^{p+1} f(a_1, \ldots, a_p) a_{p+1} + \sum_j (-1)^j f(a_1, \ldots, a_j a_{j+1}, \ldots, a_{p+1})$$

for $p > 0$, and $\delta^0 m(a) = am - ma$. Let $H^p(A, M)$ denote the vector space $\text{Ker } \delta^p / \text{Im } \delta^{p-1}$.

Now let X and Y be left A-modules and let $L(Y, X)$ denote the space of complex linear maps from Y to X. There is a natural A-bimodule structure on $L(Y, X)$ given by $(a\ell)(y) = a(\ell(y))$ and $(\ell a)(y) = \ell(ay)$ for $\ell \in L(Y, X)$, $a \in A$, $y \in Y$. The Hochschild cohomology of A with coefficients in the bimodule $L(Y, X)$ has a particularly useful interpretation.

4.2. PROPOSITION ([5], chapter III and X). Let

$$(4.2) \qquad \ldots \to P_2 \to P_1 \to P_0 \to Y \to 0$$

be a resolution of Y by projective left A-modules, and consider the induced cochain complex

$$(4.3) \qquad 0 \to \text{hom}_A(P_0, X) \to \text{hom}_A(P_1, X) \to \text{hom}_A(P_2, X) \to \ldots .$$

The cohomology of this complex is isomorphic to the Hochschild cohomology, $\dot{H}(A, L(X, Y))$, of A with coefficients in $L(X, Y)$.

We usually denote the pth cohomology group of the complex (4.3) by $\text{Ext}^p(Y, X)$ (cf. [5], Chapter III). Note that $\text{Ext}^0(Y, X) = \text{hom}_A(Y, X)$.

The functor Ext has the following properties (cf. [5], Chapter III):

(1) $\text{Ext}(Y, X) = 0$ for all X if and only if Y is projective;

(2) $\text{Ext}(Y, X) = 0$ for all Y if and only if X is injective;

(3) if $0 \to X_1 \to X_2 \to X_3 \to 0$ is a short exact sequence of left A-modules then there is a long exact sequence $0 \to \text{hom}_A(Y, X_1) \to \text{hom}_A(Y, X_2) \to \to \text{hom}_A(Y, X_3) \to \text{Ext}^1(Y, X_1) \to \text{Ext}^1(Y, X_2) \to \ldots$;

(4) if $0 \to Y_1 \to Y_2 \to Y_3 \to 0$ is a short exact sequence of left A-modules, then there is a long exact sequence, $0 \to \text{hom}_A(Y_3, X) \to \text{hom}_A(Y_2, X) \to \to \text{hom}_A(Y_1, X) \to \text{Ext}^1(Y_3, X) \to \ldots$.

The relationship of the above to our work here is as follows. Let $a = (a_1, \ldots, a_n)$ be a commuting n-tuple of operators on a Banach space X. We let A be the polynomial algebra on n-generators over $\mathbb{C}$ and consider X to be an A-module by letting these generators act as the operators $a_1, \ldots, a_n$. If $z \in \mathbb{C}^n$, then $\mathbb{C}_z$ will denote the A-module consisting of $\mathbb{C}$ with the generators of A acting as multiplication by the coordinates $z_1, \ldots, z_n$ of z. We then have the following proposition:

4.3. PROPOSITION. For each p, the cohomology group $H^p(a, X)$ of Definition 1.1 is isomorphic to $\mathrm{Ext}^p(\mathbb{C}_z, X)$. Therefore, $z \in Sp(a, X)$ if and only if $\mathrm{Ext}^p(\mathbb{C}_z, X)$ fails to be zero for some p.

The proof amounts to noting that the complex $F(a, X)$ of Definition 1.1 is of the form (4.3), where the projective resolution (4.2) is given by the Koszul resolution for the polynomial algebra A with augmentation $p \to p(z)$ (cf. [5], Chapter VII).

Now the construction of the complex $F(X, a)$ in §1 and the ensuing definition of nonsingularity and spectrum depended very heavily on the commutativity of the tuple a. However, the definition of Ext not only does not require that A be a polynomial algebra, it does not require that A be a commutative. We submit, therefore, that an adaptation of the ideas of homological algebra to analysis may yield a noncommutative version of spectral theory, including an analytic functional calculus for functions of "noncommuting variables". We are currently investigating this possibility -- with confusion and dismay the main by products to date.

REFERENCES

1. R. Arens, The analytic functional calculus in topological algebras, Pac. J. Math., 11 (1961), pp. 405-429.
2. R. Arens and A. P. Calderon, Analytic functions of several Banach algebra elements, Ann. of Math., 62 (1955), pp. 204-216.
3. N. Dunford, and J. T. Schwartz, Linear Operators, part II, Interscience, 1963.
4. R. C. Gunning and H. Rossi, Analytic Functions of Several Complex Variables, Prentice-Hall, 1965.
5. S. MacLane, Homology, Springer-Verlag, 1963.
6. S. MacLane and G. Birkoff, Algebra, MacMillan, 1967.
7. G. E. Shilov, On the decomposition of a normed ring into a direct sum of ideals, Amer. Math. Soc. Transl. (2), 1 (1955), pp. 37-48.
8. J. L. Taylor, A joint spectrum for several commuting operators, J. of Fnl. Anal., to appear.
9. ____________, The analytic functional calculus for several commuting operators, submitted to Acta Math.

REPRESENTATION OF CERTAIN OPERATORS IN HILBERT SPACE

G. K. Kalisch – *University of California, Irvine*

This paper contains several representation theorems of abstractly given bounded operators on a Hilbert space $\mathcal{H}$ as concrete multiplication, integration, or substitution operators on $\mathcal{L}_2$ spaces. The well-known prototype of these (Theorem 1) says that hermitian operators on $\mathcal{H}$ are representable as multiplication operators on $\mathcal{L}_2$ spaces. The method of proof that we use is described in the introductory part of this paper; for more detail see [2]. The substance of Theorem 3 is classical. It deals with the simultaneous representation of finite commuting jointly cyclic sets of hermitian operators as multiplication operators by the various independent variables on $\mathcal{L}_2(m)$, where m is a Borel measure with compact support in $\mathbb{R}^N$. Theorem 5 adds a necessary and sufficient condition that insures that the measure m factor into the product of one-dimensional Borel measures. Theorem 2 on the characterization of the simple Volterra operator (see [2]) and Theorem 4 form a pair; the latter is a characterization of direct sums or direct integrals of simple Volterra operators. Theorem 6 contains a characterization of certain substitution operators. The various remarks made about unitary invariants will be pursued systematically elsewhere.

The techniques used in this paper were used by B. N. Harvey [1] in his recent thesis to obtain concrete representations of operators on a Hilbert space that has also defined on it an indefinite metric (or indefinite inner product) given by a bounded hermitian operator J such that J^2 equals identity; the corresponding indefinite inner product ((x, y)) is given by ((x, y)) = (Jx, y). Such Hilbert spaces are called ***J-spaces***; we allow the two eigenspaces of J to be infinite dimensional. The representation theorems concern ***J-hermitian*** and ***J-unitary*** operators, i.e., operators T such that $T^* = JTJ$ and $T^* = JT^{-1}J$ respectively. The representations sought are implemented, not by isometries of Hilbert spaces, but by bounded and boundedly invertible linear operators between two J-spaces that preserve the indefinite inner products. The representing J-spaces are $\mathcal{L}_2$ spaces. His results include various chacterizations of certain cyclic J-hermitian and J-unitary operators as multiplication and integration operators on suitable $\mathcal{L}_2$ spaces.

We shall use the following conventions. All operators will be ***bounded***; two operators S and T acting in the Hilbert spaces $\mathcal{H}$ and $\mathcal{K}$ respectively will be called ***isomorphic*** if there exists an isometry U of $\mathcal{H}$ onto $\mathcal{K}$ such that TU = US; two sets $\{S_j\}$ and $\{T_j\}$ will be called ***isomorphic*** or ***simultaneously isomorphic*** if $T_jU = US_j$ for all j. We also say that T or $\{T_j\}$ on $\mathcal{K}$ affords a ***representation*** or ***simultaneous representation*** of S or $\{S_j\}$ on $\mathcal{H}$. We say that S on $\mathcal{H}$ is ***cyclic*** if there exists in $\mathcal{H}$ an element e called ***cyclic element*** so that the linear combinations of e, Se, $S^2e, \ldots$

are dense in $\mathcal{H}$. A set of operators acting in $\mathcal{H}$ is called *cyclic* (or *jointly cyclic*) if there exists a *cyclic element* e in $\mathcal{H}$ such that the least closed subspace containing e that is invariant under the operators in the set is $\mathcal{H}$. As usual, we denote the real numbers by $\mathbb{R}$, the nonnegative real numbers by $\mathbb{R}_+$ and the complex numbers by $\mathbb{C}$. All our $\mathcal{L}_2$ spaces are $\mathcal{L}_2(m)$ with Borel measure m with compact support in $\mathbb{R}$ or $\mathbb{R}^N$ as the case may be. We write $\mathcal{L}_2(0,1)$ for Lebesgue measure. We write V for the simple Volterra operator defined by $(Vf)(t) = \int_0^x f(y)dy$. If $e \in \mathcal{H}$, we write $\{e\}$ for the one-dimensional subspace of $\mathcal{H}$ generated by e. I is the unit interval in $\mathbb{R}$, I^N is the unit interval in $\mathbb{R}^N$. We write M_φ for the operator defined by $(M_\varphi f)(t) = \varphi(t)f(t)$; M_t is defined by $(M_t f)(t) = tf(t)$; we write $M = M_t$. We write κ and ρ both for complex and real numbers and for the operators $\kappa \cdot$ Identity and $\rho \cdot$ Identity in $\mathcal{H}$.

We start by stating, and indicating proofs of, several well-known theorems, most of which are classical.

THEOREM 1. Every cyclic hermitian operator T on the Hilbert space $\mathcal{H}$ is isomorphic with the operator M on a suitable $\mathcal{L}_2(m)$ space.

The proof here is based on verifying and exploiting the equation

$$(T^a e, T^b e) = (M^a u, M^b u) \tag{1}$$

for all nonnegative integers a and b; e is the cyclic element for T, and the function u identically equal to one is cyclic for the operator M on $\mathcal{L}_2(m)$. The real Borel measure m on $\mathbb{R}$ may be determined, and equation (1) verified, by checking that a theorem of Herglotz type on analytic functions mapping the upper half-plane into itself is applicable:

$$((T - z)^{-1} e, e) = \int (t - z)^{-1} dm(t). \tag{2}$$

The right side of this equation equals $((M - z)^{-1} u, u)$, defined on the Hilbert space $\mathcal{L}_2(m)$, and we obtain (1) by equating coefficients; see [2] for details. Another way of proving the theorem is to replace T by an operator $S = \rho T + \sigma$ for suitable real $\rho \neq 0$ and σ so that $0 \leqslant S \leqslant 1$. The representation of S as M on $\mathcal{L}_2(m)$ can then be used to obtain a representation of T. We check that the Hausdorff Moment Theorem applies and implies that

$$(S^a e, e) = \int_0^1 t^a dm(t) \tag{3}$$

for all nonnegative integers a. The right side of this equation equals $(M^a u, u)$ defined on the Hilbert space $\mathcal{L}_2(m)$ and we thus obtain (1) again. The isometry U implementing the isomorphism of T on $\mathcal{H}$ with M on $\mathcal{L}_2(m)$ is determined by

$$U \Sigma \kappa_j T^j e = \Sigma \kappa_j M^j u \tag{4}$$

for all complex numbers κ_j and all finite sums as discussed in detail in [2]; we have the equation

$$(5) \qquad UT = MU.$$

Note incidentally that the uniqueness of the measure m in (2) and (3) implies that the unitary invariant of cyclic hermitian T is the equivalence class of measures determined by m.

The situation for cyclic unitary operators T is exactly the same; here of course T on its Hilbert space $\mathcal{H}$ is isomorphic with $M_{e^{it}}$ on a suitable $\mathcal{L}_2(m)$. Equation (1) must now be verified for all integers a and b. The basis for this is either Herglotz' original theorem implying that

$$(6) \qquad ((T + z)(T - z)^{-1}e, e) = \int_0^{2\pi} ((e^{it} + z)/(e^{it} - z))dm(t)$$

or the Trigonometric Moment Theorem implying that

$$(7) \qquad (T^a e, e) = \int_0^{2\pi} e^{iat} dm(t)$$

for all integers a. The proof again proceeds as outlined above in equations (1), (4) and (5). As in the previous case, the uniqueness of the measure m in (6) and (7) implies that the unitary invariant of cyclic unitary T is the equivalence class of measures determined by m.

For Volterra operators we have the following theorem.

THEOREM 2. Let T on $\mathcal{H}$ be cyclic with cyclic element e. Suppose that $(T + T^*)\mathcal{H} = \{e\}$ and that the spectrum of T consists of zero alone. Then T on $\mathcal{H}$ is isomorphic with ρV on $\mathcal{L}_2(0,1)$ where the non-zero real number ρ is determined by $(T + T^*)e = \rho e$.

See [2] for details; the proof proceeds by establishing (1) by first showing that

$$1 - \rho((T - z^{-1})^{-1}e, e) = e^{\rho z},$$

where the right side also equals $1 - \rho((\rho V - z^{-1})^{-1}u, u)$, and then equating coefficients in a related equation.

There are theorems stating that finite commuting sets of hermitian, Volterra, and unitary operators may be simultaneously represented by multiplication or integration operators, In the hermitian and unitary cases much stronger theorems are available and well-known; the proofs indicated here for the jointly cyclic case are new, suggest an extension of Herglotz type theorems to several complex variables, and allow the formulation of simultaneous unitary invariants for the sets of operators in question. We shall present here only the hermitian case; the unitary case is similar and the more complicated Volterra case may be found in [3].

THEOREM 3. Every finite jointly cyclic set of N commuting hermitian operators $\{T_j\}_{j=1}^N$ on the Hilbert space $\mathcal{H}$ is isomorphic with the set $\{M_j\}_{j=1}^N$ on a suitable $\mathcal{L}_2(m)$ space where the support of m is in $\mathbb{R}^N$ and where

$$(M_j f)(t_1, \ldots, t_n) = t_j f(t_1, \ldots, t_n).$$

The proof proceeds just as in the case of a single operator; (1) is replaced by

$$(8) \qquad (T_1^{a_1} \ldots T_N^{a_N} e, T_1^{b_1} \ldots T_N^{b_N} e) = (M_1^{a_1} \ldots M_N^{a_N} u, M_1^{b_1} \ldots M_N^{b_N} u).$$

for all nonnegative integers $a_1, \ldots, b_N$. The procedure used for a single hermitian operator called either for a Herglotz type theorem or for the Hausdorff Moment Theorem. The first alternative suggests a generalization to several complex variables of this Herglotz type theorem and will be presented elsewhere. The second alternative may be based on the N-dimensional version of the Hausdorff Moment Theorem [5]. We must first replace $\{T_j\}$ by $\{S_j\}$ where $S_j = \rho_j T_j + \sigma_j$ with real $\rho_j \neq 0$ and σ_j such that $0 \leqslant S_j \leqslant 1$. We then use the proposition [6] that the product of finitely many commuting hermitian positive operators is again positive and are then in a position to verify that the hypotheses of the N-dimensional Hausdorff Moment Theorem are satisfied so that

$$(9) \qquad (S_1^{a_1} \ldots S_N^{a_N} e, e) = \int_0^1 \ldots \int_0^1 t_1^{a_1} \ldots t_N^{a_N} dm(t).$$

We then proceed exactly as in the case of a single hermitian operator as outlined above. Just as in the case of a single hermitian operator, the uniqueness of the measure m in (9) implies that the joint unitary invariant of a finite commuting jointly cyclic set of hermitian operators is the equivalence class of measures determined by m.

THEOREM 4. Let the operator T, defined on the separable Hilbert space $\mathcal{H}$, have spectrum zero. Define $\mathcal{E}$ as the closure of $(T + T^*)\mathcal{H}$ and suppose that the subspace $(T;\mathcal{E})$ of $\mathcal{H}$ generated by T and E is all of $\mathcal{H}$. Suppose further that if two subspaces $\mathcal{E}_1$ and $\mathcal{E}_2$ of $\mathcal{E}$ are orthogonal, so are the subspaces $(T;\mathcal{E}_1)$ and $(T;\mathcal{E}_2)$ of $\mathcal{H}$ generated by T and $\mathcal{E}_1$ and $\mathcal{E}_2$, respectively. Then T on $\mathcal{H}$ is isomorphic with $M_{\varphi(y)}V_x$ defined on $\mathcal{L}_2(m)$, where the support of the Borel measure m lies in the unit square of $\mathbb{R}^2$, m is the product of Lebesgue measure (on the x-axis) and a suitable Borel measure n (on the y-axis), and

$$((M_{\varphi(y)}V_x)f)(x, y) = \varphi(y) \int_0^x f(s, y)ds.$$

Remark. The conditions on T are of course satisfied by $M_{\varphi(y)}V_x$; the crucial hypotheses are that the spectrum of T consist of zero alone and the orthogonality of $(T;\mathcal{E}_1)$ and $(T;\mathcal{E}_2)$ for orthogonal $\mathcal{E}_1$ and $\mathcal{E}_2$; we shall call this the ***orthogonality condition.***

For the proof, denote by h the restriction $T + T^*|_{\mathcal{E}}$ of $T + T^*$ to $\mathcal{E}$, and let h be the bounded measurable function $\varphi(k)$ of the cyclic hermitian operator k on $\mathcal{E}$ with cyclic element $e \in \mathcal{E}$. Let $f(\cdot)$ be the resolution of the identity for k so that $k = \int \lambda df(\lambda)$ and $h = \varphi(k) = \int \varphi(\lambda)df(\lambda)$. Define a resolution of the identity $F(\cdot)$ on $\mathcal{H}$ by

$$F(\Gamma) \Sigma \kappa_j T^j f(\Delta_j)e = \Sigma \kappa_j T^j f(\Delta_j \cap \Gamma)e,$$

where the Borel sets $\Delta_j \subset \mathbb{R}$ are disjoint; the κ_j are complex numbers and the sums are finite; $\Gamma \subset \mathbb{R}$ is an arbitrary Borel set; e is a cyclic element for k in $\mathcal{E}$. The function $F(\cdot)$ is well defined by the orthogonality condition and is densely defined since $(T;\mathcal{E}) = \mathcal{H}$ and k is cyclic. The idempotent character of F is clear and its hermitian nature is a consequence of the orthogonality condition. It suffices to show that for three Borel sets Δ_1, Δ_2, and Γ and any two integers r and s we have

$$(F(\Gamma)T^r f(\Delta_1)e, T^s f(\Delta_2)e) = (T^r f(\Delta_1)e, F(\Gamma)T^s f(\Delta_2)e).$$

But this follows from the definition of F and the orthogonality condition. If $\Delta = \Delta_1 \cap \Delta_2 \cap \Gamma$, we see that both sides of the equation equal $(T^r f(\Delta)e, T^s f(\Delta)e)$. Let now $K = \int \lambda dF(\lambda)$ and $H = \varphi(K) = \int \varphi(\lambda)dF(\lambda)$. Let P be the orthogonal projection onto $\mathcal{E}$. It is not hard to check that $TF = FT$. We also need to know that P commutes with F. Note first that $F = f$ on $\mathcal{E}$. Next observe that if $e^\perp \in \mathcal{E}^\perp$, then $PFe^\perp = 0$, for, if $x \in \mathcal{H}$ then $(PFe^\perp, x) = (e^\perp, FPx) = 0$ since $FPx = fPx \in \mathcal{E}$. Therefore, if $x \in \mathcal{H}$ is written as $x = e' + e^\perp$ with $e' \in \mathcal{E}$ and $e^\perp \in \mathcal{E}^\perp$, then we have $PFx = PF(e' + e^\perp) = PFe' = Pfe' = fe'$ and $FPx = FP(e' + e^\perp) = FPe' = fe'$ also. Now use the direct integral representation $k = \int \lambda I(\mathcal{H}_\lambda)dn(\lambda)$, $\mathcal{H} = \int \mathcal{H}_\lambda dn(\lambda)$, where n is a Borel measure on the unit interval of R and where $I(\mathcal{H}_\lambda)$ is the identity operator on $\mathcal{H}_\lambda$; see for example [4]. We remark that there is no loss of generality if we do take n to have its support on the unit interval; this amounts to choosing k to have its spectrum in the unit interval. The commutation relations $FT = TF$ and $PF = FP$ imply that $T = \int T_\lambda dn(\lambda)$, $T^* = \int T_\lambda{}^* dn(\lambda)$ and $P = \int P_\lambda dn(\lambda)$ where T_λ and P_λ are defined on $\mathcal{H}_\lambda$. Write $\mathcal{E}_\lambda = P_\lambda \mathcal{H}_\lambda$. We then have $\mathcal{E} = \int \mathcal{E}_\lambda dn(\lambda)$, $k = \int \lambda I(\mathcal{E}_\lambda)dn(\lambda)$, $h = \int \varphi(\lambda)I(\mathcal{E}_\lambda)dn(\lambda)$. Since k is cyclic, we must have $\dim \mathcal{E}_\lambda \leqslant 1$, We also have $\mathcal{E}_\lambda \neq 0$ a.e., i.e., $P_\lambda \neq 0$ a.e. Otherwise $P_\lambda = 0$ for some set Λ of positive measure. Let $x = \int x_\lambda dn(\lambda)$ be different from zero and such that $x_\lambda \neq 0$ for $\lambda \in \Lambda$ and $x_\lambda = 0$ for $\lambda \notin \Lambda$. If now Λ were a set of full n-measure, we would have $P = 0$ and so $\mathcal{E} = 0$, contradicting $(T;\mathcal{E}) = \mathcal{H}$. Otherwise we have $x \perp (T;\mathcal{E})$ still contradicting $(T;\mathcal{E}) = \mathcal{H}$ since $(T^j Py, x) = \int_\Lambda (T^j_\lambda P_\lambda y_\lambda, x_\lambda)dn(\lambda) = 0$ for all $y = \int y_\lambda dn(\lambda)$ in $\mathcal{H}$. Thus $\dim \mathcal{E}_\lambda = 1$ a.e. There is no loss of generality in altering $e = \int e_\lambda dn(\lambda)$ so that $\|e_\lambda\| = 1$ for all λ (we may choose n to have finite total mass). Our next aim is to show that a.e. T_λ on $\mathcal{H}_\lambda$ satisfies the hypotheses of Theorem 2. First T_λ on $\mathcal{H}_\lambda$ is cyclic with cyclic element e_λ. This is a consequence of the condition $(T;\mathcal{E}) = \mathcal{H}$. Second, $(T_\lambda + T_\lambda{}^*)\mathcal{H}_\lambda = \{e_\lambda\}$. This results from the equation $\overline{(T + T^*)\mathcal{H}} = \mathcal{E}$. Third, the spectrum of T_λ is zero a.e. This is a consequence of our hypothesis on T saying that its spectrum consists of zero alone. Thus

T_λ on $\mathcal{H}_\lambda$ is isomorphic with $\psi(\lambda)V$ for real non-zero $\psi(\lambda)$ on $\mathcal{L}_2(0,1)$; in particular $(T_\lambda^r e_\lambda, T_\lambda^s e_\lambda) = \psi(\lambda)^{r+s}(V^r u, V^s u)$ where u is the function identically equal to one and r and s are non-zero integers. We shall write $T_\lambda = \psi(\lambda)V_\lambda$ where $V_\lambda: \mathcal{H}_\lambda \to \mathcal{H}_\lambda$ corresponds to V. We then have $T + T^* = \int \psi(\lambda)(V_\lambda + V_\lambda^*)dn(\lambda)$ and $T + T^*|_{\mathcal{E}} = \int \psi(\lambda)I(\mathcal{E}_\lambda)dn(\lambda) = \int \varphi(\lambda)I(\mathcal{E}_\lambda)dn(\lambda)$ and so $\varphi = \psi$ a.e. The isomorphism of T on $\mathcal{H}$ with $M_{\varphi(y)}V_x$ on $\mathcal{L}_2$(n x Lebesgue) is implemented by the isometry U of $\mathcal{H}$ onto $\mathcal{L}_2$ determined by

$$U \Sigma \kappa_j T^j f(\Gamma_j)e = \Sigma \kappa_j M^j \varphi(y)\chi_{\Gamma_j}(y)V_x^j u$$

for all complex κ_j, all nonnegative integers j, all finite sums and all disjoint Borel sets $\Gamma_j \subset R$; χ_Γ is the characteristic function of Γ. The function U is well-defined by the orthogonality condition on T; the equation $UT = M_{\varphi(y)}V_x U$ is a consequence of the definition; the isometric character of U results from the equations $(T^r f(\Gamma)e, T^s f(\Delta)e) = \int \varphi(\lambda)^{r+s}(V^r u, V^{s} u)dr(\lambda) = (M^r{}_{\varphi(y)}\chi_\Gamma(y)V_x^r u, M^s{}_{\varphi(y)}\chi_\Gamma(y)V_x^s u)$.

As in connection with the preceding representation theorems, we remark that two operators as described in the present theorem are unitarily equivalent if and only if their corresponding M_φ's are unitarily equivalent.

THEOREM 5. Let $\{T_j\}_{j=1}^N$ be a finite commutative jointly cyclic set of hermitian operators on the Hilbert space $\mathcal{H}$ such that if $F_j(\cdot)$ is the resolution of the identity of T_j and $\{\Gamma_j\}_{j=1}^N$ is a family of Borel sets in $\mathbb{R}$, then

$$\Pi_j F_j(\Gamma_j) = 0 \text{ implies at least one of } F_j(\Gamma_j) = 0.$$

Then $\{T_j\}$ on $\mathcal{H}$ is isomorphic with $\{M_j\}$ on $\mathcal{L}_2(m)$, where the support of m lies in the product of N intervals in $\mathbb{R}^N$ and m factors into N one-dimensional Borel measures along the N coordinate axes of R^N; M_j is given by

$$(M_j f)(t_1, \ldots, t_N) = t_j f(t_1, \ldots, t_N).$$

Alternatively, there exist Hilbert spaces $\{\mathcal{H}_j\}_{j=1}^N$ and cyclic hermitian operators $\{h_j\}_{j=1}^N$ defined on them such that $\mathcal{H} = \bigotimes_{j=1}^N \mathcal{H}_j$ and $T_j = \bigotimes_{j=1}^N h'_k$ where h'_k is the identity on $\mathcal{H}_k$ for $k \neq j$ and $h'_j = h_j$. Thus T_j has uniform multiplicity equal to the product of the dimensions of the $\mathcal{H}_k$ for $k \neq j$, and $\Pi T_j = \otimes h_j$.

Remarks. The corresponding theorem for finite commuting sets of hermitian matrices (of finite size) may be formulated with the same conclusions if the hypothesis about the resolutions of the identity is replaced by demanding that if ρ_j is an eigenvalue of T_j $(j = 1, \ldots, N)$ then there exists a single joint non-zero eigenvector v such that for all $j = 1, \ldots, N$ we have

$$T_j v = \rho_j v.$$

The proof is a simple exercise.

The situation of the present theorem corresponds to the case of Theorem 3 where the measure in $\mathbb{R}^N$ is equivalent to a product measure. It is clear that any set of operators such as occurs in the conclusion of the present theorem satisfies its hypotheses.

We present the proof for $N = 2$; the general case is proved in the same way except that many subscripts and dots must be used. Use the direct integral representation $\mathcal{H} = \int \mathcal{H}(\lambda)dm(\lambda)$ and $T_1 = H = \int \lambda I(\mathcal{H}_\lambda)dm(\lambda)$. Then since $T_2 = K$ commutes with H, we have $K = \int K_\lambda dm(\lambda)$, where K_λ is a bounded hermitian operator acting in $\mathcal{H}_\lambda$. We also have $e = \int e_\lambda dm(\lambda)$ with $e_\lambda \in \mathcal{H}_\lambda$, where $e \in \mathcal{H}$ is cyclic for the pair H, K. We now claim 1) that K_λ is cyclic for a.a. λ with cyclic element e_λ, and 2) that there exists a cyclic operator K_0 on a Hilbert space $\mathcal{H}_0$ so that K_λ on $\mathcal{H}_\lambda$ is isomorphic with K_0 on $\mathcal{H}_0$. 1) Let the subspace $(K_\lambda;e_\lambda)$ generated by K_λ and e_λ in $\mathcal{H}_\lambda$ be denoted by $\mathcal{K}_\lambda$. We can form $\int \mathcal{K}_\lambda dm(\lambda) = \mathcal{K}$ since K_λ acts in $\mathcal{H}_\lambda$, $e_\lambda \in \mathcal{H}_\lambda$ with $e = \int e_\lambda dm(\lambda)$, and $\int K_\lambda dm(\lambda) = K$. If $\mathcal{K} = \mathcal{H}$, then since $\int \mathcal{K}_\lambda^\perp dm(\lambda)$ also exists and equals $\mathcal{K}^\perp = (0)$, we can conclude that $\mathcal{K}_\lambda^\perp = (0)$ a.e., and so $\mathcal{K}_\lambda = \mathcal{H}_\lambda$. If $\mathcal{K} \neq \mathcal{H}$, then if we take a non-zero element $k^\perp = \int k_\lambda^\perp dm(\lambda) \in \mathcal{K}^\perp$, we see that $(H^iK^je, k^\perp) = 0$, so e is not cyclic for the pair H, K, since then $(H^iK^je, k^\perp) = \int \lambda^i (K^j_\lambda e_\lambda, k_\lambda^\perp)dm(\lambda) = 0$. 2) Let $F(\cdot)$ denote the resolution of the identity of K; then let it be determined by the countable set $\{\Delta_j\}$ of Borel sets on $\mathbb{R}$. Let $F(\Delta) = \int F_\lambda(\Delta)dm(\lambda)$. If $F(\Delta_r) = 0$, then so is $F_\lambda(\Delta_r) = 0$ a.e.; if $F(\Delta_r) \neq 0$ we claim $F_\lambda(\Delta_r) \neq 0$ a.e. Otherwise suppose that there existed a Borel set Λ of positive m-measure, i.e., such that $E(\Lambda) \neq 0$, where $F_\lambda(\Delta_r) = 0$; here $E(\cdot)$ is the resolution of the identity of H. Then $E(\Lambda)F(\Delta_r) = \int_\Lambda F_\lambda(\Delta_r)dm(\lambda) = 0$, contradicting our hypothesis. Thus if we except a set of m-measure zero, we conclude that $F_\lambda(\Delta) \neq 0$ if and only if $F(\Delta) \neq 0$, and so since the operators K_λ are cyclic a.e., they are unitarily equivalent a.e. and we can write $K_\lambda = \int \mu dn(\mu)$ where n does not depend on λ, and $\mathcal{H}_\lambda = \int \mathcal{H}_{\lambda\mu}dn(\mu)$, $e_\lambda = \int e_{\lambda\mu}dn(\mu)$. As we may assume both m and n to have finite support and finite mass on $\mathbb{R}$, we may also assume without loss of generality that $e_{\lambda\mu} = 1$ -- note that $\mathcal{H}_{\lambda\mu}$ has dimension one since K_λ is cyclic. We may also assume that $\|e_\lambda\| = 1$. We now calculate as follows: $(H^iK^je, H^rK^se) = \int \lambda^{i+r}(K_\lambda^{j+s}e_\lambda, e_\lambda)dm(\lambda) = \int \lambda^{i+r} \int \mu^{j+s}dn(\mu)dm(\lambda) = (M_x^iM_y^ju, M_x^rM_y^su)$ on $\mathcal{L}_2(m \times n)$ where u is the function identically equal to one. Define now

$$U \Sigma \kappa_{ij} H^iK^je = \Sigma \kappa_{ij} M_x^iM_y^ju$$

on $\mathcal{L}_2 = \mathcal{L}_2(m \times n)$. The function U is well-defined for all finite sums and all complex numbers κ_{ij} and is an isometry. Since the terms on the left are dense in $\mathcal{H}$ by hypothesis and those the the right are dense in $\mathcal{L}_2$ by a well-known theorem, the function U extends to an isometry of $\mathcal{H}$ onto $\mathcal{L}_2$ so that $UH = M_xU$ and $UK = M_yU$.

The conclusions of the theorem can now be read off from the various assertions proved above. For example, $\mathcal{L}_2(m \times n) = \mathcal{H}_1 \otimes \mathcal{H}_2$ where $\mathcal{H}_1 = \mathcal{L}_2(m)$ and $\mathcal{H}_2 = \mathcal{L}_2(n)$; $T_1 = H = \int H_\lambda dm(\lambda)$ and $T_2 = K = \int K_\lambda dm(\lambda)$ with the K_λ a.e. unitarily equivalent implies that $\dim \mathcal{H}_\lambda$ will be constant a.e. and therefore H has uniform multiplicity equal to the dimension of $\mathcal{L}_2(n) = \mathcal{H}_2$. The remainder of the statements may be read off from the simultaneous isomorphism of the pair H, K on $\mathcal{H}$ with the pair M_x, M_y on $\mathcal{L}_2$.

We remark that the simultaneous unitary equivalence of the sets $\{H_j\}_{j=1}^N$ and $\{K_j\}_{j=1}^N$ of commuting hermitian operators satisfying the hypotheses of the theorem is equivalent to their being pairwise and separately unitarily equivalent.

THEOREM 6. Suppose the invertible linear operator T on $\mathcal{H}$ satisfies

$$(TT^*)(T^*T) = (T^*T)(TT^*) \tag{10}$$

and that TT^* is cyclic. Then T on $\mathcal{H}$ is isomorphic with $M_F M_\psi U_\psi$ on a suitable $\mathcal{L}_2(m)$ where m is a Borel measure on $\mathbb{R}$, F and ψ are m-bounded and m-measurable functions with ψ nonnegative and $|F| = 1$ a.e., and the unitary operator $U_\psi = M_v S_\psi$ with $v \in \mathcal{L}_2(m)$ and with S_ψ the substitution operator $(S_\psi f)(t) = f(\psi(t))$.

Remarks. 1) For invertible matrices T, equation (10) implies that T is unitarily equivalent to the product of matrices $D_p VP$ where D_p is diagonal with positive entries, V is unitary and commutes with D_p and P is a permutation matrix; if TT^* is cyclic, V is diagonal and unitary. There are obvious modifications if T is not invertible.

2) The hypothesis of invertibility is not essential; if we can represent $T + \rho$ for suitable real ρ then T itself is also representable.

3) The converse of the theorem is true. If $S = M_F M_\psi U_\psi$ as described in the theorem, then $SS^* = M_\psi^2$ and $S^*S = M_t^2$ so that (10) is true.

In order to prove the theorem, represent simultaneously $(TT^*)^{1/2}$ as M_ψ and $(T^*T)^{1/2}$ as M_t on a suitable $\mathcal{L}_2(m)$ space where the support of the Borel measure m lies on $\mathbb{R}_+$; the function ψ is nonnegative a.e. Since M_t and M_ψ are unitarily equivalent, there exists a unitary operator mapping $\mathcal{L}_2(m)$ onto itself such that $UM_t = M_\psi U$. As the function u identically equal to one is cyclic for M_t, $Uu = v$ is cyclic for M_ψ. We have $(M_t^r u, u) = (U^* M_\psi^r Uu, u) = (M_\psi^r v, v)$ or

$$\int \psi(t)^r |v(t)|^2 dm(t) = \int t^r dm(t). \tag{11}$$

Define the substitution operator S_ψ by $(S_\psi f)(t) = f(\psi(t))$; it is certainly defined for all polynomials and so is at least densely defined. The operator $U_\psi = M_v S_\psi$ is an isometry of a dense subset of $\mathcal{L}_2(m)$ into itself, for, if f is a polynomial then $\|U_\psi f\| = \|f\|$; this follows from equation (11): If $f = \Sigma \kappa_j t^j$ then $\|U_\psi f\|^2 = \|M_v f(\psi(t))\|^2 = \int |v(t)|^2 |\Sigma \kappa_j \psi(t)^j|^2 dm(t) = \Sigma \kappa_r \bar{\kappa}_s \int \psi(t)^{r+s} |v(t)|^2 dm(t) = \Sigma \kappa_r \bar{\kappa}_s \int t^{r+s} dm(t) = \|f\|^2$. We next show that U_ψ maps $\mathcal{L}_2(m)$ onto a dense set. Since by hypothesis M_t and so M_ψ are cyclic with cyclic function u and $Uu = v$ respectively, we know that the sets $\{M_t^r u\}$ and $\{M_\psi^r v\}$ are linearly dense. Therefore the set $\{U_\psi t^r\}$ is linearly dense since $U_\psi t^r = v(t)\psi(t)^r = M_\psi^r v$. Thus U_ψ may be extended to an isometry of $\mathcal{L}_2(m)$ onto itself. We have

$$U_\psi M_t = M_\psi U_\psi \tag{12}$$

since $U_\psi M_t t^r = v(t)\psi(t)^{r+1}$ and $M_\psi U_\psi t^r = \psi(t)v(t)\psi(t)^r$. Since T is invertible we have as above the unitary operator U such that $M_\psi U = UM_t = T$; this together with (12) implies that $U^{-1}M_\psi U = U_\psi^{-1}M_\psi U_\psi$ or $UU_\psi^{-1}M_\psi = M_\psi UU_\psi^{-1}$. The cyclicity of M_ψ implies now that there exists an m-measurable function F of absolute value one so that $UU_\psi^{-1} = M_F$, i.e., so that $U = M_F U_\psi$, and finally $T = M_F M_\psi U_\psi$ as desired. We have used the same letter T for the original operator on $\mathcal{H}$ as for its representation on $\mathcal{L}_2(m)$. Unitary invariants of the operators here under discussion are more complicated and will be discussed elsewhere.

REFERENCES

1. B. N. Harvey, Representation of certain linear operators in Hilbert space, Thesis, 1969, Univ. Calif. Irvine.
2. G. K. Kalisch, Direct proofs of spectral representation theorems, J. Math. Anal. Appl., 8, 1964, 351 - 363.
3. G. K. Kalisch, Characterizations of direct sums and commuting sets of Volterra operators, Pac. J. Math., 18, 1966, 545 - 552.
4. M. A. Naimark and S. V. Fomin, Continuous direct sums of Hilbert spaces, Usp. Mat. Nauk, 10, 2 (64), 1955, 111 - 143 = Amer. Math. Soc. Transl., (2) 5, 1957, 35 - 66.
5. J. A. Shohat and J. D. Tamarkin, Problem of moments, Amer. Math. Soc., New York, 1943.
6. B. Sz.-Nagy, Spektralderstellung Hermitescher Operatoren, Berlin, 1942.

SOME INEQUALITIES INVOLVING THE HARDY-LITTLEWOOD MAXIMAL FUNCTION IN A THEORY OF CAPACITIES

C. Preston – *University of California, San Diego*

I. INTRODUCTION. In Section II we briefly outline a theory of capacities for the circle group T (i.e. $\{z: |z| = 1\}$) as developed by Carleson in [1]. We start with a kernel $K \in \mathcal{L}_1(T)$ satisfying certain conditions and define a capacity theory with respect to K. At the end of the section we introduce a new kernel $\overline{K}$, related to K (and as defined by Carleson), and find out under what conditions $\overline{K}$ also satisfies the conditions that K does.

In Section III we obtain some inequalities involving the Hardy-Littlewood maximal function that extend those given by Carleson in [1] and [2]. We use a norm $\|f\|_K$ for $f \in \mathcal{L}_2(T)$ given by $\|f\|_K^2 = \Sigma |\hat{f}(n)|^2 \lambda_n$, where $\lambda_n = [\hat{K}(n)]^{-1}$. Denoting by θ_f the Hardy-Littlewood maximal function of a nonnegative function f in $\mathcal{L}_1(T)$ we prove that $\int \theta_f(x) d\sigma(x) \leqslant 4\pi \|f\|_K [\iint K(x - y) d\sigma(x) d\sigma(y)]^{1/2}$ for any nonnegative Borel measure σ. We also give an elementary proof of an inequality that appeared in [3], namely that if K is such that $K(x) = O(K(2x))$ as $x \to 0$ then we have

$$C_K(\{x: \theta_f(x) > \lambda\}) \leqslant \pi_K \|f\|_K^2/\lambda^2,$$

where π_K is a constant depending only on K.

The author would like to thank Professor Adriano M. Garsia for his advice and guidance on the writing of this paper.

II. AN OUTLINE OF A THEORY OF CAPACITIES. We consider a kernel $K: T \to \mathbb{R}$ having the properties:

(i) K is continuous;

(ii) $K \geqslant 0$;

(iii) K is symmetric (i.e. $K(t) = K(2\pi - t)$);

(iv) K is decreasing (i.e. $K(t) \leqslant K(s)$ for $0 \leqslant s \leqslant t \leqslant \pi$);

(v) K is convex;

(vi) $K \in \mathcal{L}_1(T)$.

We will almost always have $K(0) = +\infty$.

It is well-known that a function satisfying the above conditions is also positive-definite.

Let $\lambda_n = [\hat{K}(n)]^{-1}$; then $0 < \lambda_0 \leq \lambda_n$, and $\lambda_n = \lambda_{-n}$. We consider the Hilbert space $\mathcal{H}_K$ where

$$\mathcal{H}_K = \{f \in \mathcal{L}_2(T): \sum_{n=-\infty}^{\infty} |f(n)|^2 \lambda_n < \infty\},$$

with norm given by $\|f\|_K = [\sum_{n=-\infty}^{\infty} |f(n)|^2 \lambda_n]^{1/2}$.

Given $\mu, \nu \in M^+(T)$ (the nonnegative Borel measures on T) we define $u_\mu = K * \mu$, (the potential associated with μ); and $I_K(\mu,\nu) = \int K * \mu \, d\nu$ $(= \iint K(x-y) \, d\mu(x) d\nu(y))$.

THEOREM 1. Suppose $\mu_n \to \mu$, $\nu_n \to \nu$ weakly, where $\mu_n, \mu, \nu_n, \nu \in M^+(T)$. Then

(a) $\varliminf_{n\to\infty} I_K(\mu_n, \nu_n) \geq I_K(\mu, \nu)$;

(b) $\varliminf_{n\to\infty} u_{\mu_n}(x) \geq u_\mu(x)$;

(c) $\varliminf_{x_n \to x_0} u_\mu(x_n) \geq u_\mu(x_0)$.

The proof of this theorem is simple, and is given in [3].

THEOREM 2. If $\mu \in M^+(T)$ and $u_\mu(x) \leq 1$ for $x \in \operatorname{supp}(\mu)$ then $u_\mu(x) \leq 1$ for all $x \in T$.

Proof. Clearly u_μ is continuous in $T - \operatorname{supp}(\mu)$. If we write $T - \operatorname{supp}(\mu) = \bigcup_{j=1}^{\infty} I_j$ as the disjoint union of the open intervals I_j, then the fact that K is convex implies that u_μ is sublinear on each I_j. The theorem thus follows from the following easily proved lemma.

LEMMA. Let u_μ be as in the theorem, and write $T - \operatorname{supp}(\mu) = \bigcup_{j=1}^{\infty} I_j$ as before. Let (a,b) be the interval I_j for some j. Then $u_\mu(a) = \lim_{x \downarrow a} u_\mu(x)$ and $u_\mu(b) = \lim_{x \uparrow b} u_\mu(x)$.

Let $E \subset T$; we define $U_E \subset M^+(T)$ by $U_E = \{\mu \in M^+(T): \operatorname{supp}(\mu) \subset E$, and $u_\mu(x) \leq 1$ for $x \in E\}$.

We define the *capacity* $C_K(E)$ of E with respect to K by

$$C_K(E) = \sup\{\|\mu\|: \mu \in U_E\}.$$

If a property holds except on a set E, with $C_K(E) = 0$, we say it holds C_K-almost everywhere (C_K-a.e.).

We note that if $E \subset T$ is a Borel set and $\nu \in M^+(T)$ is such that u_ν is bounded on E and $C_K(E) = 0$, then $\nu(E) = 0$. In particular if $C_K(E) = 0$ then E has Lebesgue measure zero.

THEOREM 3. We have:

(a) C_K is an increasing set function;

(b) $C_K(E) = \sup\{C_K(F)$: F is compact and $F \subset E\}$;

(c) C_K is a countably subadditive set function, i.e.

$$C_K\left(\bigcup_{n=1}^{\infty} E_n\right) \leqslant \sum_{n=1}^{\infty} C_K(E_n);$$

(d) C_K is translation invariant.

The proof of Theorem 3 is immediate.

THEOREM 4. Let $F \subset T$ be compact. Then there exists $\mu \in U_F$ such that

(a) $u_\mu(x) = 1$ C_K-a.e. on F;

(b) $\mu(F) = C_K(F)$.

We say that μ is the *equilibrium measure*, and u_μ is the *equilibrium potential* for F. A proof of this theorem can be found in [1] or [3].

If we have a kernel K satisfying the properties (i), . . . , (vi) above, we can define a new kernel $\overline{K}$ by:

$$\overline{K}(x) = \left(\int_0^{|x|} K(t)dt\right)/|x| \quad (\text{for } |x| \leqslant \pi).$$

We have the following result:

THEOREM 5. Let K be a kernel satisfying (i), . . . , (vi). Then $\overline{K}$ satisfies (i), . . . , (v). Also $\overline{K} \geqslant K$, and $\overline{K}$ satisfies (vi) (i.e. $\overline{K} \in \mathcal{L}_1(T)$ if and only if $H \in \mathcal{L}_1(T)$, where $H(x) = \log(K(x)/|x|)$.

Proof. It is clear that $\overline{K}$ satisfies (i), (ii) and (iii), and also that $\overline{K} \geqslant K$. Since $\overline{K}'(x) = K(x)/x - (\int_0^x K(t)dt)/x^2 = (K(x) - \overline{K}(x))/x \leqslant 0$, $\overline{K}$ is decreasing.

We show $\overline{K}$ is convex by proving that $\overline{K}'(x)$ is an increasing function of x (for $0 \leqslant x \leqslant \pi$). Take $0 < s < t \leqslant \pi$. We have

$$\overline{K}'(t) - \overline{K}'(s) = (K(t) - \overline{K}(t))/t - (K(s) - \overline{K}(s))/s.$$

For $0 < r \leqslant \pi$ define A_r to be the area of the region bounded by the lines $x = 0$, $y = K(r)$, and the curve $y = K(x)$. Then $A_r = r(\overline{K}(r) - K(r))$ Therefore $\overline{K}'(t) - \overline{K}'(s) = A_s/s^2 - A_t/t^2 = (t^2 A_s - s^2 A_t)/s^2 t^2 = ((t^2 - s^2)A_s - s^2(A_t - A_s))/s^2 t^2$. From figure 1 (page 29), $A_s \geqslant$ area $\Delta BCE = s^2(\tan\theta)/2$. Also $A_t - A_s \leqslant$ area $EFDC = (t - s)(t + s)(\tan\theta)/2$. Thus $(t^2 - s^2)A_s \geqslant s^2(A_t - A_s)$ and so $\overline{K}'(t) - \overline{K}'(s) \geqslant 0$. Therefore $\overline{K}$ is convex.

The fact that $\overline{K} \in \mathcal{L}_1(T)$ if and only if $H \in \mathcal{L}_1(T)$ follows easily on using Fubini's theorem.

III. SOME INEQUALITIES INVOLVING THE HARDY-LITTLEWOOD MAXIMAL FUNCTION. Let $f \geqslant 0$ with $f \in \mathcal{L}_1(T)$. We define the (right) Hardy-Littlewood maximal function of f, denoted θ_f, by

$$\theta_f(x) = \sup\{(\int_x^{x+h} f(t)dt)/h:\ h > 0\}.$$

We define the two-sided maximal function φ_f by

$$\varphi_f(x) = \sup\{(\int_{x-h}^{x+h} f(t)dt)/2h:\ h > 0\}.$$

Let $D = \{z: |z| < 1\}$. Given $f \in \mathcal{L}_1(T)$ we define P_f to be the harmonic function associated with f, i.e. $P_f\colon D \to \mathbb{C}$ can be given by $P_f(re^{it}) = \sum_{n=-\infty}^{\infty} r^{|n|}\hat{f}(n)e^{int}$.

Note that if $f \geqslant 0$, then $P_f \geqslant 0$.

It is well-known that if $f \geqslant 0$ and $f \in \mathcal{L}_1(T)$ then

$$\theta_f(x)/4\pi \leqslant \sup\{P_f(re^{ix}):\ r < 1\} \leqslant \varphi_f(x) \tag{1}$$

for all $x \in T$.

THEOREM 6. Suppose $f \in \mathcal{H}_K$ with $f \geqslant 0$. Let $\mu \in M^+(D)$. Then

$$(\int P_f(z)d\mu(z))^2 \leqslant \|f\|_K^2 \int\int \overline{K}(t_1 - t_2)d\mu(z_1)d\mu(z_2)$$

where $z_1 = r_1 e^{it_1}$, $z_2 = r_2 e^{it_2}$.

Proof. We first consider μ with $\mathrm{supp}(\mu) \subset \{z: |z| \leqslant r_0\}$ where $r_0 < 1$.

$$\begin{aligned}|\int \sum_{n=-N}^{N} r^{|n|}\hat{f}(n)e^{int}d\mu(z)|^2 &= |\sum_{n=-N}^{N} \hat{f}(n) \int r^{|n|}e^{int}d\mu(z)|^2 \\ &\leqslant (\sum_{n=-N}^{N} |\hat{f}(n)|^2\lambda_n) \sum_{n=-N}^{N} |\int r^{|n|}e^{int}d\mu(z)|^2/\lambda_n \\ &\leqslant \|f\|_K^2 \int\int \sum_{n=-N}^{N} r_1^{|n|}r_2^{|n|}\hat{K}(n)e^{in(t_1-t_2)}d\mu(z_1)d\mu(z_2).\end{aligned}$$

Since both sums converge uniformly on $\mathrm{supp}(\mu)$ we can let $N \to \infty$ to give

$$(\int P_f(z)d\mu(z))^2 \leqslant \|f\|_K^2 \iint \sum_{n=-\infty}^{\infty} r_1^{|n|} r_2^{|n|} \hat{K}(n) e^{in(t_1-t_2)} d\mu(z_1)d\mu(z_2) =$$

$$= \|f\|_K^2 \iint P_K(r_1 r_2 e^{i(t_1-t_2)}) d\mu(z_1)d\mu(z_2).$$

However from (1) we have $P_K(r_1 r_2 e^{i(t_1-t_2)}) \leqslant \varphi_K(t_1 - t_2)$; so

$$(\int P_f(z)d\mu(z))^2 \leqslant \|f\|_K^2 \iint \varphi_K(t_1 - t_2)d\mu(z_1)d\mu(z_2).$$

But we have (assume $x > 0$):

$$\varphi_K(x) = \sup\{(\int_{x-h}^{x+h} K(t)dt)/2h\colon h > 0\} \leqslant \sup\{(\int_{x-h}^{x} K(t)dt)/h\colon h > 0\} =$$

$$= (\int_0^x K(t)dt)/x = \overline{K}(x)$$

Thus for all $x \in T$ we must have $\varphi_K(x) \leqslant \overline{K}(x)$; which gives

$$(\int P_f(z)d\mu(z))^2 \leqslant \|f\|_K^2 \iint \overline{K}(t_1 - t_2)d\mu(z_1)d\mu(z_2).$$

Finally, the restriction on $\mathrm{supp}(\mu)$ is easily removed by use of the monotone convergence theorem.

We now use (1) again and Theorem 6 to prove:

THEOREM 7. Suppose $f \in \mathcal{H}_K$ with $f \geqslant 0$. Let $\sigma \in M^+(T)$. Then

$$\int \theta_f(x)d\sigma(x) \leqslant 4\pi \|f\|_K [I_{\overline{K}}(\sigma, \sigma)]^{1/2}.$$

Proof. Let $g\colon T \to (0, 1)$ be measurable. Define $G \subset D$ by $G = \{re^{it}\colon g(t) = r\}$. We project σ onto G to give $\sigma_g \in M^+(D)$ defined by $\sigma_g(E) = \sigma\{t\colon g(t)e^{it} \in E\}$, for any Borel set $E \subset D$. Now given $\epsilon > 0$, we define g by: $g(t) = s$, where s is such that $0 < s < 1$ and

$$P_f(se^{it}) > \sup\{P_f(re^{it})\colon r < 1\} - \epsilon.$$

Since P_f is continuous, g can clearly be chosen to be measurable. Using (1) we now have

$$\int P_f(z)d\sigma_g(z) = \int P_f(g(t)e^{it})d\sigma_g(z) =$$

$$= \int P_f(g(t)e^{it})d\sigma(t) \geqslant \int (\theta_f(t)/4\pi - \epsilon)d\sigma(t).$$

Therefore from Theorem 6 we have (provided $\int (\theta_f(t)/4\pi - \epsilon)d\sigma(t) \geqslant 0$) that

$$(\int (\theta_f(t)/4\pi - \epsilon)d\sigma(t))^2 \leqslant \|f\|_K^2 \iint \overline{K}(t_1 - t_2)d\sigma_g(z_1)d\sigma_g(z_2) =$$

$$= \|f\|_K^2 \iint \overline{K}(t_1 - t_2)d\sigma(t_1)d\sigma(t_2) = \|f\|_K^2 I_{\overline{K}}(\sigma, \sigma).$$

Since $\epsilon > 0$ is arbitrary we have

$$(\int \theta_f(t)d\sigma(t))/4\pi)^2 \leqslant \|f\|_K^2 I_{\overline{K}}(\sigma, \sigma),$$

which proves the theorem.

We now consider some special kernels which we denote K_a, $0 \leqslant a < 1$, where

$$K_a(x) = |x|^{-a}, \; 0 < a < 1,$$

$$K_0(x) = \log(2\pi/|x|).$$

We clearly have $\overline{K}_a(x) = K_a(x)/(1 - a)$, $0 < a < 1$, and $\overline{K}_0(x) = K_0(x) + 1/2\pi$. Thus for these kernels Theorem 7 becomes:

THEOREM 8. For $0 \leqslant a < 1$ there exists a constant B_a such that if $f \in \mathcal{H}_{K_a}$ with $f \geqslant 0$, and if $\sigma \in M^+(T)$ then

$$\int \theta_f(x)d\sigma(x) \leqslant B_a \|f\|_{K_a} [I_{K_a}(\sigma, \sigma)]^{1/2}.$$

Carleson proves this result in [2] using specific properties of the kernels K_a.
From Theorem 7 the following is easy to prove:

THEOREM 9. Let $f \in \mathcal{H}_K$ with $f \geqslant 0$. Then for any $\lambda > 0$ we have

$$C_{\overline{K}}(\{x: \theta_f(x) > \lambda\}) \leqslant 16\pi^2 \|f\|_K^2/\lambda^2.$$

Proof. Let F be any compact set with $F \subset \{x: \theta_f(x) > \lambda\}$. Let σ be the equilibrium measure for F (with respect to $\overline{K}$). Thus $\text{supp}(\sigma) \subset F$, $\sigma \in U_F$, $u_\sigma = 1$ $C_{\overline{K}}$-a.e. on F, and $C_{\overline{K}}(F) = \sigma(F)$. Therefore $I_{\overline{K}}(\sigma, \sigma) = \int u_\sigma d\sigma = \sigma(F) = C_{\overline{K}}(F)$. Also $\int \theta_f(x)d\sigma(x) \geqslant \lambda\sigma(F) = \lambda C_{\overline{K}}(F)$. Thus from Theorem 7 we have

$$\lambda C_{\overline{K}}(F) \leqslant 4\pi \|f\|_K [C_{\overline{K}}(F)]^{1/2};$$

i.e.

$$C_{\overline{K}}(F) \leqslant 16\pi^2 \|f\|_K^2/\lambda^2.$$

The theorem now follows immediately from Theorem 3 (b).

The inequality in Theorem 9 can in general be improved. In fact we will now show that if we assume that $K(x) = O(K(2x))$ as $x \to 0$, then the following is true: for all $f \in \mathcal{H}_K$ with $f \geq 0$, and for all $\lambda > 0$ we have

$$(2) \qquad C_K(\{x: \theta_f(x) > \lambda\}) \leq \pi_K \|f\|_K^2/\lambda^2,$$

where π_K is a constant depending only on K. [A proof of this appears in [3], but here we will give a proof by more elementary means.]

We will make use of the following property of the maximal function (which can easily be proved using the Sunrise Lemma of F. Riesz): Let $f \in \mathcal{L}_1(T)$ with $f \geq 0$ and let $E = \{x: \theta_f(x) > 1\}$. E is open, so write $E = \bigcup_{j=1}^{\infty} I_j$ as the disjoint union of the open intervals I_j. Then $\int_{I_j} f(t)dt = |I_j|$ for each j, (where $|I_j|$ denotes the length of the interval I_j).

Important for establishing (2) is the following theorem:

THEOREM 10. Let K be a kernel as in Section II, and let $I_1, \ldots, I_N$ be disjoint open subintervals of T. We define $B_{ij} = \int_{I_i}\int_{I_j} K(x - y)dxdy$. Suppose we have constants $b_j \geq 0$, $j = 1, \ldots, N$. Then there exist constants $c_j \geq 0$, $j = 1, \ldots, N$, such that

(a) $\sum_{i=1}^{N} c_i B_{ij} \geq b_j$, $j=1, \ldots, N$;

(b) $\sum_{j=1}^{N} c_j b_j = \sum_{j=1}^{N} \sum_{i=1}^{N} c_i c_j B_{ij}$.

We give the proof of this theorem at the end of the Section.

THEOREM 11. Let $I_1, \ldots, I_N$ be disjoint open sub-intervals of T. Consider those $f \in \mathcal{H}_K$ satisfying

$$(3) \qquad \int_{I_j} f(t)dt \geq |I_j|, \ j = 1, \ldots, N.$$

Then there exists $u \in \mathcal{H}_K$ satisfying (3) such that if $f \in \mathcal{H}_K$ satisfies (3) then we have $\|u\|_K \leq \|f\|_K$. Also u has the form $u = K * u_0$, where $u_0 = \sum_{j=1}^{N} c_j \chi_{I_j}$ for some constants $c_j \geq 0$.

Proof. We apply Theorem 10 with $b_j = |I_j|$ to obtain constants $c_j \geq 0$ such that:

(a) $\sum_{i=1} c_j B_{ij} \geq |I_j|$, $j = 1, \ldots, N$;

(b) $\sum_{j=1}^{N} c_j |I_j| = \sum_{j=1}^{N} \sum_{i=1}^{N} c_i c_j B_{ij}$.

Let $u_0 = \sum_{j=1}^{N} c_j \chi_{I_j}$, and $u = K * u_0$. Since $u_0 \in \mathcal{L}_2(T)$ it is clear that $u \in \mathcal{H}_K$; also $\|u\|_K^2 = \sum_{n=-\infty}^{\infty} |\hat{u}(n)|^2 \lambda_n = \sum_{n=-\infty}^{\infty} |\hat{u}_0(n)\hat{K}(n)|^2 \lambda_n = \sum_{n=-\infty}^{\infty} |\hat{u}_0(n)|^2 \hat{K}(n) = \sum_{n=-\infty}^{\infty} \hat{u}(n)\overline{\hat{u}_0(n)}$. Therefore since u, $u_0 \in \mathcal{L}_2(T)$ we have that

$$\|u\|_K^2 = \int u(t)u_0(t)dt = \sum_{j=1}^{N}\sum_{i=1}^{N} c_i c_j \int_{I_j} K * \chi_{I_i}(t)dt = \sum_{j=1}^{N}\sum_{i=1}^{N} c_i c_j B_{ij}.$$

Hence from (b) we have $\|u\|_K^2 = \sum_{j=1}^{N} c_j |I_j| = \|u_0\|_1$. We also have

$$\int_{I_j} u(t)dt = \sum_{i=1}^{N} c_i \int_{I_j} K * \chi_{I_i}(t)dt = \sum_{i=1}^{N} c_i B_{ij} \geqslant |I_j| \text{ (from (a))}.$$

Thus u satisfies (3). Now take $f \in \mathcal{H}_K$ satisfying (3). We calculate $(u, f)_K$ (i.e. the inner product of u and f in $\mathcal{H}_K$).

$$\begin{aligned}(u, f)_K &= \sum_{n=-\infty}^{\infty} \hat{u}(n)\overline{\hat{f}(n)}\lambda_n = \sum_{n=-\infty}^{\infty} \hat{u}_0(n)\overline{\hat{f}(n)} = \int f(t)u_0(t)dt \quad (\text{since } f, u_0 \in \mathcal{L}_2(T))\\ &= \sum_{j=1}^{N} c_j \int_{I_j} f(t)dt \geqslant \sum_{j=1}^{N} c_j |I_j| \quad (\text{since } f \text{ satisfies (3)})\\ &= \|u_0\|_1 = \|u\|_K^2.\end{aligned}$$

Therefore $\|u\|_K^2 \leqslant (u, f)_K \leqslant \|u\|_K \|f\|_K$, and so we have $\|u\| \leqslant \|f\|_K$ as was required.

We now investigate the properties of the function u obtained in Theorem 11. We will use the notation that if I is an open subinterval of T then $\tilde{I}$ is the open interval which has the same midpoint as I, and has length three times that of I.

LEMMA. Let $I_1, \ldots, I_N$ and u, u_0 be as in Theorem 11. Suppose that $\tilde{I}_1, \ldots, \tilde{I}_N$ are disjoint. Then if the kernel K is such that $K(x) \leqslant aK(2x)$ for $0 \leqslant x \leqslant \pi/2$ we have

$$\max\{u(x): x \in I_j\} \leqslant \beta \min\{u(x): x \in I_j\}, \; j = 1, \ldots, N,$$

where β $\max\{a, 2\}$.

Proof. Clearly it is sufficient to show that

$$\max\{K * \chi_{I_i}(x): x \in I_j\} \leqslant \beta \min\{K * \chi_{I_i}(x): x \in I_j\}, \; i, j = 1, \ldots, N.$$

First consider the case when $i \neq j$: if x_1, $x_2 \in I_j$, $y \in I_i$ then $|x_1 - y| \geqslant |x_2 - y|/2$. since $\tilde{I}_i \cap \tilde{I}_j = \emptyset$, and so $K(x_1 - y) \leqslant aK(x_2 - y)$. Hence $\max\{K * \chi_{I_i}(x): x \in I_j\} \leqslant a \min\{K * \chi_{I_i}(x): x \in I_j\}$. Now if $i = j$ then we have $\max\{K * \chi_{I_j}(x): x \in I_j\} = 2\int^{|I_j|/2} K(t)dt \leqslant 2\int^{|I_j|} K(t)dt = 2\min\{K * \chi_{I_j}(x): x \in I_j\}$. Hence we do have $\max\{u(x): x \in I_j\} \leqslant \beta \min\{u(x): x \in I_j\}$.

LEMMA. Let $I_1, \ldots, I_N$ be open subintervals of T such that $\tilde{I}_1, \ldots, \tilde{I}_N$ are disjoint. Suppose K is such that $K(x) \leqslant aK(2x)$ for $0 \leqslant x \leqslant \pi/2$. Then

$$C_K(\bigcup_{j=1}^{N} \tilde{I}_j) \leqslant 3a^2 C_K(\bigcup_{j=1}^{N} I_j).$$

A proof of this lemma is given in [3].

We are now in a position to verify that (2) holds.

THEOREM 12. Suppose K is such that $K(x) \leqslant aK(2x)$ for $0 \leqslant a \leqslant \pi/2$. Then for all $f \in \mathcal{H}_K$ with $f \geqslant 0$, and for all $\lambda > 0$ we have

$$C_K(\{x: \theta_f(x) > \lambda\}) \leqslant \pi_K \|f\|_K^2/\lambda^2,$$

where π_K is a constant depending only on K (and we can take $\pi_K = 9a^2 \max\{a, 2\}$).

Proof. This is essentially the same as in the proof of Theorem 18 in [3] but we will repeat it here for the sake of completeness. Let $E = \{x: \theta_f(x) > 1\}$. Write $E = \bigcup_{j=1}^{\infty} I_j$ as the disjoint union of the open intervals I_j. As previously noted we have

$$\int_{I_j} f(t)dt = |I_j| \text{ for all } j. \tag{4}$$

Let $E_M = \bigcup_{j=1}^{M} I_j$, and $E'_M = \bigcup_{j=1}^{M} \tilde{I}_j$. Let $\tilde{I}_{j_1}, \ldots, \tilde{I}_{j_s}$ be a minimal subset of $\tilde{I}_1, \ldots, \tilde{I}_M$ having the property $E'_M = \bigcup_{i=1}^{s} \tilde{I}_{j_i}$. We choose the indices $j_1, \ldots, j_s$ so that $\tilde{I}_{j_1}, \ldots, \tilde{I}_{j_s}$ is a counter clockwise ordering of the intervals. Let $H_o = \cup\{\tilde{I}_{j_i}: 3 \leqslant i \leqslant s, i \text{ odd}\}$, $H_e = \cup\{\tilde{I}_{j_i}: 2 \leqslant i \leqslant s, i \text{ even}\}$. Then $E'_M = H_o \cup H_e \cup \tilde{I}_{j_1}$, so

$$C_K(E'_M) \leqslant C_K(H_o) + C_K(H_e) + C_K(\tilde{I}_{j_1}).$$

Thus the capacity of one of H_o, H_e, $\tilde{I}_{j_1}$ must be greater than or equal to $C_K(E'_M)/3$. Without loss of generality we will suppose that $C_K(H_o) \geqslant C_K(E'_M)/3$. But the intervals comprising H_o are disjoint, so we have constructed a subcollection $G_1, \ldots, G_N$ of $I_1, \ldots, I_M$ such that $\tilde{G}_1, \ldots, \tilde{G}_N$ are disjoint and $C_K(\bigcup_{j=1}^{N} \tilde{G}_j) \geqslant C_K(\bigcup_{j=1}^{M} I_j)/3$. Thus we have $C_K(\bigcup_{j=1}^{M} I_j) \leqslant C_K(\bigcup_{j=1}^{M} \tilde{I}_j) \leqslant C_K(\bigcup_{j=1}^{N} \tilde{G}_j)/3 \leqslant 9a^2 C_K(\bigcup_{j=1}^{N} G_j)$ (by the previous lemma).

Let u be the function found in Theorem 12 corresponding to the intervals $G_1, \ldots, G_N$. Then from (4) we have $\|f\|_K \geqslant \|u\|_K$. We have $\int_{G_j} u(t)dt \geqslant |G_j|$, $j = 1, \ldots, N$, since u satisfies (3), and so $\max\{u(x): x \in \tilde{G}_j\} \geqslant 1$, $j = 1, \ldots, N$. However since $\tilde{G}_1, \ldots, \tilde{G}_N$ are disjoint we have by the first Lemma that $u(x) \geqslant 1/\beta$ on $\bigcup_{j=1}^{N} G_j$. Thus $\|u\|_K^2 = \|u_0\|_1 \geqslant C_K(\bigcup_{j=1}^{N} G_j)/\beta$. Thus $C_K(E_M) = C_K(\bigcup_{j=1}^{M} I_j) \leqslant 9a^2 C_K(\bigcup_{j=1}^{N} G_j) \leqslant$

$\leqslant 9a^2\beta\|u\|_K^2 \leqslant 9a^2\beta\|f\|_K^2$. Now let F be any compact subset of E. Then for some M we have $F \subset E_M$ and so $C_K(F) \leqslant 9a^2\beta\|f\|_K^2$. Thus from Theorem 3 we have $C_K(E) \leqslant$ $\leqslant 9a^2\beta\|f\|_K^2$. Finally, on replacing f by f/λ we obtain

$$C_K(\{x: \theta_f(x) > \lambda\}) \leqslant 9a^2\beta\|f\|_K^2/\lambda^2.$$

As proved in [3], Theorem 12 can be used to show that if $f \in \mathcal{H}_K$ then the Abel sums of f converge C_K-a.e.

We now return to the proof of Theorem 10. Recall that we are trying to find $c_j \geqslant 0$ such that

(a) $\sum_{i=1}^{N} c_i B_{ij} \geqslant b_j$, $j = 1, \ldots, N$,

(b) $\sum_{j=1}^{N} c_j b_j = \sum_{j=1}^{N} \sum_{i=1}^{N} c_i c_j B_{ij}$

Let B denote the matrix B_{ij}. Note that B is strictly positive definite (since K is positive definite, and $\iint K(x - y)g(x)g(y)dxdy = 0$ for $g \in \mathcal{L}_\infty(T)$ implies $g = 0$ a.e.). Thus in particular B is non-singular. We will use the notation that $x \geqslant y$ for $x = (x_1, \ldots, x_N) \in \mathbb{R}^N$, $y = (y_1, \ldots, y_N) \in \mathbb{R}^N$, if $x_j \geqslant y_j$, $j = 1, \ldots, N$; we will write $b = (b_1, \ldots, b_N)$. Thus what we are trying to find is $c \in \mathbb{R}^N$ such that $c \geqslant 0$, $cB \geqslant b$ and $cb^T = cBc^T$.

We can define a strictly positive quadratic form $Q: \mathbb{R}^N \times \mathbb{R}^N \to \mathbb{R}$ by $Q(x,y) = xBy^T$. Let $S \subset \mathbb{R}^N$ be given by $S = \{x \in \mathbb{R}^N: xB \geqslant b\}$. S is a closed, convex subset of $\mathbb{R}^N$. There thus exists $c \in S$ such that $Q(c,c) = \min\{Q(x,x): x \in S\}$. Let $x \in S$; then $(1 - \delta)c + \delta x \in S$ for $0 \leqslant \delta \leqslant 1$. Hence $Q(c,c) \leqslant Q((1 - \delta)c + \delta x, (1 - \delta)c + \delta x) = Q(c,c) + 2\delta Q(x,c) - 2\delta Q(c,c) + \delta^2 Q(c - x, c - x)$; but δ can be made arbitrarily small so we must have $Q(c,c) \leqslant Q(x,c)$. Hence $c \in S$ is characterized by the property that $Q(c,c) \leqslant Q(x,c)$ for all $x \in S$. Now take $a \in \mathbb{R}^N$ such that $a \geqslant 0$. Since B is non-singular there exists $y \in \mathbb{R}^N$ such that $yB = a$. Then $(c + y)B \geqslant b + a \geqslant b$, so $c + y \in S$ and thus $Q(c,c) \leqslant Q(c + y, c) = Q(c,c) + Q(y,c)$. Hence $0 \leqslant$ $\leqslant Q(y,c) = yBc^T = ac^T$. Therefore $ac^I \geqslant 0$ for all $a \geqslant 0$ and so we must have $c \geqslant 0$.

We will now show that $cb^T = cBc^T$, and thus c will solve our problem. Since $cB \geqslant b$ and $c \geqslant 0$ we have $cBc^T \geqslant bc^T = cb^T$. On the other hand, since B is non-singular there exists $y \in R^N$ such that $yB = b$. Then $y \in S$ and so $cBc^T = Q(c,c) \leqslant$ $\leqslant Q(y,c) = yBc^T = bc^T = cb^T$. Thus $cBc^T = cb^T$.

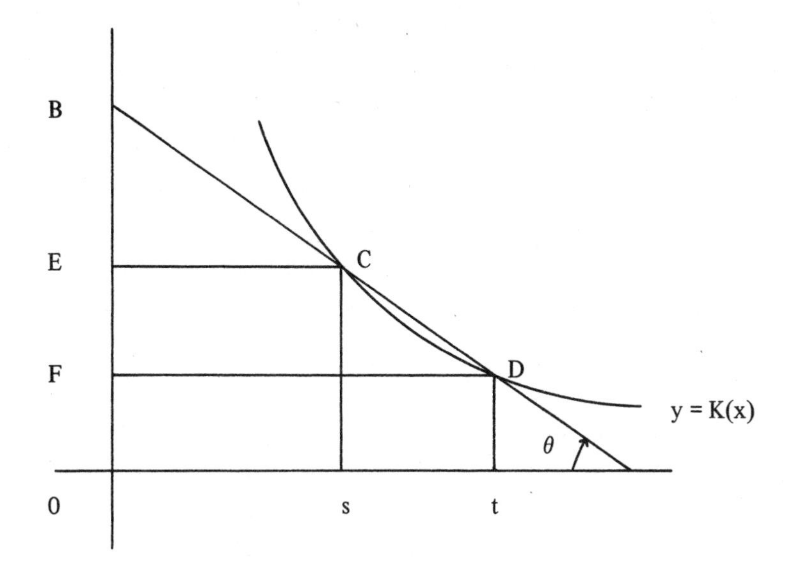

Figure 1

REFERENCES

1. L. Carleson, Selected Problems on Exceptional Sets, Van Nostrand Mathematical Studies #13.
2. L. Carleson, Maximal functions and capacities, Ann. Inst. Fourier, 15 (1965), 59-64.
3. C. Preston, A theory of capacities and its application to some convergence results, (to appear).

A DECOMPOSITION THEOREM FOR AUTOMORPHISMS OF VON NEUMANN ALGEBRAS

Robert R. Kallman[1] – *Yale University*

In the 1930's von Neumann defined the notion of free action and applied it to the construction of certain factors. Given an abelian von Neumann algebra $\mathcal{R}$ and a $*$-automorphism φ of $\mathcal{R}$, φ is said to be *freely acting* on $\mathcal{R}$, if, given a nonzero projection P in $\mathcal{R}$, there exists a nonzero projection Q in $\mathcal{R}$ satisfying

$$Q \leqslant P \quad \text{and} \quad Q \perp \varphi(Q).$$

Now let $\mathcal{R}$ be an arbitrary abelian von Neumann algebra and let φ be an arbitrary $*$-automorphism of $\mathcal{R}$. Let $P_2 = \sup\{P\colon P$ is a projection in $\mathcal{R}$ and, given any projection Q in $\mathcal{R}$ such that $Q \leqslant P$, we have $\varphi(Q) = Q\}$. Let $P_1 = I - P_2$. One may easily check that $\mathcal{R} = \mathcal{R}_{P_1} \oplus \mathcal{R}_{P_2}$, $\varphi(\mathcal{R}_{P_i}) \subset \mathcal{R}_{P_i}$ $(i = 1, 2)$, φ is freely acting on $\mathcal{R}_{P_1}$, and φ is the identity on $\mathcal{R}_{P_2}$.

In this lecture we will discuss a (non-trivial) non-commutative generalization of this decomposition. We first generalize the notion of free action. Let $\mathcal{R}$ be an arbitrary von Neumann algebra and φ is a $*$-automorphism of $\mathcal{R}$. Then φ is said to be *freely acting* on $\mathcal{R}$ if the only A in $\mathcal{R}$ which satisfies $AB = \varphi(B)A$ $(B \in \mathcal{R})$ is $A = 0$. It is not at all evident that this definition of free action coincides with von Neumann's for abelian $\mathcal{R}$. We sketch the proof of this. Suppose $\mathcal{R}$ is abelian, that φ is freely acting in von Neumann's sense, and that $BA = AB = \varphi(B)A$ $(B \in \mathcal{R})$. Let P be the range projection of A. Suppose A is nonzero; then P is also nonzero, and one has that $BP = \varphi(B)P$ $(B \in \mathcal{R})$. Choose a nonzero projection Q such that $Q \leqslant P$ and $Q \perp \varphi(Q)$. Then $Q = Q^2 = Q^2P = Q\varphi(Q)P = 0$. Contradiction. Hence A is zero. Conversely, suppose that P is a nonzero projection in $\mathcal{R}$. We want to find a nonzero projection $Q \leqslant P$ such that $Q \perp \varphi(Q)$. If no such Q exists, then $\varphi(Q) = Q$ for all $Q \leqslant P$. To show this, first suppose $Q \not\leqslant \varphi(Q)$. Let $Q' = Q - Q\varphi(Q)$. Then $0 \neq Q' \leqslant P$ and $\varphi(Q') \leqslant \varphi(Q) \perp Q'$. Next, suppose $\varphi^{-1}(Q) < Q$. Let $Q' = Q - \varphi^{-1}(Q)$. Then $0 \neq Q' \leqslant \leqslant P$ and $Q' \leqslant Q \perp \varphi(Q')$. Hence $\varphi(Q) = Q$ for all $Q \leqslant P$. Thus if Q is a projection in $\mathcal{R}$, then $\varphi(Q)P = \varphi(QP) = QP$. Therefore $PB = BP = \varphi(B)P$ $(B \in \mathcal{R})$. Hence $P = 0$. Contradiction. Thus there exists a nonzero projection $Q \leqslant P$ such that $Q \perp \varphi(Q)$. Hence the two definitions coincide in case $\mathcal{R}$ is abelian.

The main result is the following theorem.

---

[1] Most of the topics covered in this talk are taken from a paper by the author which is scheduled to appear shortly in the Duke Mathematical Journal.

THEOREM 1. Let $\mathcal{R}$ be any von Neumann algebra and φ any $*$-automorphism of $\mathcal{R}$. Then $\mathcal{R} = \mathcal{R}_1 \oplus \mathcal{R}_2$, $\varphi = \varphi_1 \oplus \varphi_2$, $\varphi_i(\mathcal{R}_i) \subset \mathcal{R}_i$ $(i = 1, 2)$, φ_1 is inner on $\mathcal{R}_1$, and φ_2 is freely acting on $\mathcal{R}_2$. This decomposition is unique.

Notice that in case $\mathcal{R}$ is a factor (i.e., if Cent $\mathcal{R}$ is the scalar field), this theorem implies that φ is outer if and only if it is freely acting.

We sketch the proof of this theorem in the following sequence of lemmas. We discuss the first, key, lemma in some detail since it is of interest in itself. Let T be in $\mathcal{R}$. Recall that $S(T) = \inf\{P: P$ is a projection in $\mathcal{R}$ and $TP = T\}$ is the support of T, and $C(T) = \inf\{Q: Q$ is a projection in Cent $\mathcal{R}$ and $TQ = T\}$ is the central support of T.

LEMMA 2. Let $\mathcal{R}$ be a von Neumann algebra and φ a $*$-automorphism of $\mathcal{R}$. Then φ is inner if and only if there exists $A \in \mathcal{R}$, with $C(A) = I$, such that $AB = \varphi(B)A$ $(B \in \mathcal{R})$.

Proof. It is clear that if φ is inner, then such an A exists. Conversely, suppose such an A exists. Then $B^*A^* = A^*\varphi(B^*)$ $(B \in \mathcal{R})$. Let B be any unitary in $\mathcal{R}$. Then $B^*A^*AB = A^*A$ and $AA^* = \varphi(B)AA^*\varphi(B^*)$. Hence $AA^* \in$ Cent $\mathcal{R}$ and $A^*A \in$ Cent $\mathcal{R}$. Hence $I = C(A) = C(A^*A) = S(A^*A) = S(|A|)$, for $T \in$ Cent $\mathcal{R}$ implies $C(T) = S(T)$. Similarly, $S(|A^*|) = I$, for $I = C(A) = C(A^*)$. Hence if $A = W|A|$ is the polar decomposition of A, then W is unitary in $\mathcal{R}$ since the initial projection of the partial isometry W is $S(|A|)$ and the final projection of W is $S(|A^*|)$. Now $A^*A \in$ Cent $\mathcal{R}$ implies $|A| \in$ Cent $\mathcal{R}$. Therefore $WB|A| = W|A|B = AB = \varphi(B)A = \varphi(B)W|A|$ $(B \in \mathcal{R})$. $S(|A|) = I$ implies $|A|$ has dense range. Hence $WB = \varphi(B)W$ $(B \in \mathcal{R})$, and φ is inner. The proof is complete.

We remark in passing that one may prove the following result in a similar manner. Let $\mathcal{R}$ be a von Neumann algebra and φ a $*$-automorphism of $\mathcal{R}$. Then φ is spatial if and only if there exists a bounded operator A, with $S(A) = S(A^*) = I$, which satisfies $AB = \varphi(B)A$ $(B \in \mathcal{R})$.

LEMMA 3. Let $\mathcal{R}$ be a von Neumann algebra and φ a $*$-automorphism of $\mathcal{R}$. Let $\{Q_a: a \in A\}$ be a collection of central projections in $\mathcal{R}$ such that $\varphi(Q_a) = Q_a$ and φ is inner on $\mathcal{R}_{Q_a}$ $(a \in A)$. Let $Q = \vee\{Q_a: a \in A\}$. Then Q is a central projection in $\mathcal{R}$, $\varphi(Q) = Q$, and φ is inner on $\mathcal{R}_Q$.

The proof is omitted.

LEMMA 4. Let $\mathcal{R}$ be a von Neumann algebra and φ a $*$-automorphism of $\mathcal{R}$. Suppose φ satisfies the following condition: given a central projection P such that $\varphi(P) = P$, then φ is outer on $\mathcal{R}_P$. Then φ is freely acting on $\mathcal{R}$.

The proof is omitted. Hint: use Lemma 2.

Proof of Theorem 1: The uniqueness of the decomposition is easy and is left to the reader. To show the existence of such a decomposition, let $A = \{P: P$ is a central projection in $\mathcal{R}$, $\varphi(P) = P$, and $\varphi|_{\mathcal{R}P}$ is inner$\}$.

Let $Q = \vee\{P: P \in A\}$. By Lemma 3, Q is a central projection, $\varphi(Q) = Q$, and φ is inner on $\mathcal{R}_Q$. φ is freely acting on $\mathcal{R}_{(I-Q)}$ by Lemma 4. Let $\mathcal{R}_1 = \mathcal{R}_Q$, $\mathcal{R}_2 = \mathcal{R}_{(I-Q)}$, $\varphi_1 = \varphi|_{\mathcal{R}_1}$, and $\varphi_2 = \varphi|_{\mathcal{R}_2}$. This proves Theorem 1.

One can prove the following corollaries of Theorem 1 with a little work.

COROLLARY 5. Let $\mathcal{R}$ and $\mathcal{S}$ be von Neumann algebras. Let φ be a $*$-automorphism of $\mathcal{R}$ and ψ a $*$-automorphism of $\mathcal{S}$, then $\varphi \otimes \psi$ acts freely on $\mathcal{R} \otimes \mathcal{S}$ if and only if either φ acts freely on $\mathcal{R}$ or ψ acts freely on $\mathcal{S}$.

COROLLARY 6. Let $\mathcal{R}$ and $\mathcal{S}$ be von Neumann algebras. Let φ be a $*$-automorphism of $\mathcal{R}$ and ψ a $*$-automorphism of $\mathcal{R}$ and ψ a $*$-automorphism of $\mathcal{S}$. Then $\varphi \otimes \psi$ is outer on $\mathcal{R} \otimes \mathcal{S}$ if and only if φ is outer on $\mathcal{R}$ or ψ is outer on $\mathcal{S}$.

In closing we remark that there seems to be a pleasant analogy between $*$-automorphisms of nonatomic abelian von Neumann algebras with a faithful, finite, normal, invariant state (i.e., measure-preserving automorphisms of nonatomic, probability measure spaces) and $*$-automorphisms of II_1 factors. For example, it is known that all the powers of an ergodic measure-preserving automorphism of a nonatomic probability measure space are freely acting. The alogous statement for II_1 factors is that all the powers of an ergodic $*$-automorphism are outer. Here is a quick proof of this fact. Suppose that $\mathcal{R}$ is a II_1 factor, that φ is an ergodic $*$-automorphism of $\mathcal{R}$, and that φ^n is inner. Then there exists some unitary operator U in $\mathcal{R}$ such that $\varphi^n(T) = UTU^*$ $(T \in \mathcal{R})$. Embed U on a maximal abelian subalgebra $\mathcal{A}$ of $\mathcal{R}$. Then $\varphi^n(T) = T$ for all T in $\mathcal{A}$. Choose a nonzero projection P in $\mathcal{A}$ with $tr(P) < 1/n$. Let $Q = P + \varphi(P) + \ldots + \varphi^{n-1}(P)$. Then $\varphi(Q) = Q$, so Q is some scalar multiple of the identity, say $Q = \lambda I$. Note that

$$0 < \lambda = tr(Q) = n\, tr(P) < 1.$$

But $\lambda P = P + \varphi(P)P + \ldots + \varphi^{n-1}(P)P$ and so $\lambda\, tr(P) = tr(P) +$ nonnegative terms. Hence, $\lambda \geqslant 1$. Contradiction. Hence, all powers of φ are outer. It remains to be seen just how far this analogy can be pushed.

RECENT PROGRESS IN THE STRUCTURE THEORY OF FACTORS

J. T. Schwartz – *Courant Institute of Mathematical Sciences*

My talk will be expository, and in part preparatory to the lecture of Professor Sakai which is to follow. Let me begin by reminding you that a W*-algebra is an algebra A of operators T in a separable Hilbert space H, such that

i. A is closed under the adjoint map $T \to T^*$;

ii. A is closed in the weak topology.

In the theory of these algebras, one is aiming at a systematic infinite dimensional generalization of the classical Wedderburn structure theory of finite dimensional rings. A first principal success comes with a theorem of von Neumann which tells us that each such algebra is, if not a direct sum of simple algebras, at least a ***direct integral*** of W*-algebras having one-dimensional centers. Direct integrals of Hilbert spaces and W*-algebras are here defined in a manner directly generalizing the usual notion of a direct sum; the inevitable technical problems concerning measurability, etc., are not hard to overcome. Readers are referred to [1], Chapter 1, for details.

The fact that a general W*-algebra can be expressed as a direct integral of W*-algebras having one-dimensional centers naturally focuses attention on this class of W*-algebras. These algebras were called ***factors*** by von Neumann and Murray, who began their structural analysis. A first example of a factor is the ring B(H) of all operators in H. More generally, we may consider a Kronecker product Hilbert space $H = H_1 \otimes H_2$ and the product ring $B(H) \otimes I$ in it. This W*-algebra is readily seen to be a factor also. A nonzero projection E in a factor A is called ***minimal*** if $E \geqslant F$ has no nonzero solution in A. The methods of the classical Wedderburn theory are easily adapted to show that a factor which contains a minimal projection is necessarily isomorphic to a factor $B(H_1) \otimes I$; factors of this elementary kind constitute the ***type I factors*** of von Neumann and Murray.

This partial structural result leads to the consideration of factors not containing minimal projections. If we make use of the mechanisms of group theory, such factors are not hard to construct. Let G be a countable discrete group. Introduce an "invariant measure" on G by assigning each element $g \in G$ the measure 1. We may then consider the Hilbert space $\mathcal{L}_2(G)$ defined as the set of all complex functions f such that

$$|f|_2^2 = \int_G |f(g)|^2 dg < \infty,$$

and in it the W*-algebra A(G) generated by the group of left translations. This algebra may be called the left W*-group algebra of G. It is easily seen that A(G) may also be regarded as the convolution algebra whose elements are all those functions $g \in \in \mathcal{L}_2$ such that there exists a finite constant K for which an inequality

$$|g * f|_2 \leqslant K|f|_2$$

is satisfied. It is readily seen that g belongs to the center of A(G) if and only if it is a *class function*, i.e. if and only if $g(y) \equiv g(x^{-1}yx)$. Thus, since $g \in \mathcal{L}_2$, it follows immediately that the following condition implies that A(G) is a factor:

Class Condition: Every set $C_{y_0} = \{x^{-1}y_0x\colon\ x \in G\}$ except C_e is infinite.

Thus every member of a wide family of groups defines a factor. To the extent that A(G) reflects the structure of G we may therefore surmise that numerous non-isomorphic factors will exist. It was in fact proved quite early that if Φ_2 is the free group on 2 generators and π the permutation group of countably many objects, then $A(\Phi_2)$ and $A(\pi)$ are non-isomorphic. In recent work of McDuff and of Professor Sakai himself an uncountable infinity of groups G_a all satisfying the class condition and having non-isomorphic algebras $A(G_a)$ has been exhibited. Since Professor Sakai will discuss this construction in some detail, I shall pass over this very interesting work.

By noting that the group algebras A(G) all share a common property not possessed by every factor, the existence of still another type of factor may be established. The bounded linear functional $tr(f) = f(e)$ may be defined for every group algebra. Since

$$tr(f * g) = \int_G f(x)g(x^{-1})dx,$$

it follows that $tr(f * g) = tr(g * f)$. A factor A not of type I and admitting a necessarily unique linear functional $tr(\cdot)$ such that $tr(TS) = tr(ST)$ for all $S, T \in A$ is said to be of *type II*. Factors which are neither of type I nor of type II are said to be of *type III*; such factors A may also be defined by the fact that if E and F are any two nonzero projections in A there exists an operator T in A whose range is EH and such that T* has range FH. Type II factors for which the trace functional is bounded (resp. unbounded) are said to be of *type II_1* (resp. *type II_∞*). A type II_∞ factor may be written as the Kronecker product of a type II_1 and a type I factor, so that the essential objects of structural investigation in the theory of factors are the type II_1 and the type III factors.

Type III factors can be obtained by generalizing the group-algebra construction used above. To do so, we let h be an auxiliary Hilbert space, G as before a countable discrete group, and U_g a unitary representation of G in h. We let B be a W*-algebra of operators in h, and suppose that $U_g B U_g^* = B$ for $g \in G$. Then we may consider the Hilbert space $H = \mathcal{L}_2(G,h)$ of h-valued functions f on G satisfying

$$|f|_2^2 = \int |f(x)|^2 dx < \infty,$$

and in this space the algebra of all bounded operators having the form

$$(1) \qquad (Tf)(x) = \int \tau(xy^{-1})f(y)dy,$$

where $\tau(x)$ is a function defined on G and with values in the ring B(h), subject to the additional restriction that $\tau(x) = b(x)U(x)$, where $b(x) \in B$ for $x \in G$. This *generalized group algebra* will be designated in what follows by the symbol A(G, B, U). Note that if h is one-dimensional and $U(X) \equiv 1$ the generalized group algebra A(G, B, U) reduces to the previously considered algebra A(G). Conditions of various degrees of generality guaranteeing that A(G, B, U) is a factor may readily be stated. For our pusposes, the following somewhat particularized result will be sufficient.

LEMMA 1 (cf. [1], Lemma II, 5.11). Let S be a space and μ a finite nonatomic measure in it, $h = \mathcal{L}_2(S,\mu)$ and B the abelian W*-algebra of all multiplication operators in h. Suppose that G acts in S as a group of transformations preserving measurability and μ-null sets, and that

i. G acts *ergodically*, i.e., any measurable set $\sigma \in S$ invariant under G is either μ-null or has a μ-null complement;

ii. G acts *freely*, i.e., if $g \neq e$, then $\mu(\{s \in S: sg = s\}) = 0$.

Let $\Delta_g(s)$ be the Radon-Nikodym derivative defined by

$$(2) \qquad \mu(\sigma g^{-1}) = \int_\sigma \Delta_g(s)\mu(ds),$$

so that

$$(3) \qquad (U_g f)(s) = f(sg)\{\Delta_g(sg)\}^{1/2}$$

defines a unitary representation of G in $h = \mathcal{L}_2(S,\mu)$. Then A(G, B, U) is a factor.

The following rather trivial case of Lemma 1 is worth noting for subsequent use. Let $S = \{p_1, \ldots, p_n\}$ be a set of n points, assigned measures $\{\lambda_1, \ldots, \lambda_n\}$ with $\sum_{i=1}^{n} \lambda_i = 1$. Let G be the group of shift operations S_k: $p_j \to p_{j-k}$ (mod n). Then, by Lemma 1, A(G,B,U) is a factor acting in the space $H = \mathcal{L}_2(G, h)$ which is clearly n^2-dimensional. We shall write $\Lambda = \{\lambda_1, \ldots, \lambda_n\}$, and designate this factor as $R(\Lambda)$. Since the multiplication algebra B in this case is n-dimensional, it is plain from (1) that $R(\Lambda)$ is n^2-dimensional. It follows at once from the known form of type I factors that H may be written as the Kronecker product $H = H_1 \otimes H_2$ of two n-dimensional spaces in such a way as to bring $R(\Lambda)$ into isomorphism with the ring $B(H_1) \otimes \otimes I$.

If in Lemma 1 G acts on S as a group of measure-preserving transformations, then we may define tr(T) for the operator T of (1) simply by writing $tr(T) = \varphi(b(e))$, where the functional φ is defined for a multiplication operator m by $\varphi(m) = = \int_S m(x)dx$. In this case it is easily verified that $tr(TS) = tr(ST)$, so that here $A(G, B, U)$ is either finite dimensional or it is a factor of type II_1. The contrary case is that in which no measure invariant under the group G is equivalent to μ. In such a case, as the following lemma shows, $A(G, B, U)$ is of type III.

LEMMA 2. Suppose, in addition to the hypotheses of Lemma 1, that there exists no nonzero, nonnegative, μ-measurable function f on S such that $f(sg)\Delta_{g^{-1}}(s) = = f(s)$ μ-almost everywhere for each $g \in G$. Then $A(G, B, U)$ is of type III.

In the situation described by Lemma 2, we designate the algebra $A(G, B, U)$ by the symbol $A_1(G, S, \mu)$.

The hypothesis of Lemma 2 is equivalent to the statement that there exist no G-invariant μ-continuous measure on S. An interesting class of examples may be constructed as follows. Let σ be the 2-point measure space with points p_1, p_2 having measures λ, $1 - \lambda$, where $\lambda/(1 - \lambda) = x$. We denote this measure ν_x. Let γ be the group of permutations of σ, so that γ contains precisely 2 elements. Let $S = \sigma \times \times \sigma \times \ldots$ be the product of infinitely many copies of σ, and let $G = \gamma \times \gamma \times \ldots$ be the direct product of infinitely many copies of γ. Let μ_x be the infinite direct product measure $\nu_x \times \nu_x \times \ldots$. The group G acts on the space S coordinate-by-coordinate, and each element of G permutes a finite number of the coordinates of S; G is a countable group. It is not hard to see that the only decently behaved G-invariant measure on S is μ_1. It follows (cf. [1], pp. 192-195) that the factor $A_1(G, S, \mu_x) = = R^{(x)}$ is of type III for every $x = 1$. The measures μ_x are inequivalent to one another and it is thus plausible that the factors $R^{(x)}$ should be structurally distinct. That this is indeed the case is the content of the following theorem of Powers.

THEOREM 3. No two of the type III factors $R^{(x)}$, $0 < x < 1$, are isomorphic.

In proving Theorem 3, it will be convenient to make use of the notion of an infinite Kronecker product, for a systematic account of which [2] may be consulted. Let $\{H_n\}$ be a sequence of Hilbert spaces, and $\{x_n\}$ a sequence of unit vectors with $x_n \in H_n$. The map $y \to y \otimes x_{n+1}$ defines an isometric imbedding of $H_1 \otimes \ldots \otimes H_n$ into $H_1 \otimes \ldots \otimes H_n \otimes H_{n+1}$, and thus the sequence $\{x_n\}$ defines a sequence of isometric imbeddings

$$H_1 \otimes \ldots \otimes H_n \to H_1 \otimes \ldots \otimes H_{n+1}.$$

The direct limit of this sequence of Hilbert spaces may be written as

$$H = \otimes_{n=1}^{\infty, \{x_n\}} H_n$$

and is called the infinite Kronecker product relative to the sequence $\{x_n\}$ of vectors. If $y_n \in H_n$ and $\sum_{n=1}^{\infty} |1 - (x_n, y_n)| < \infty$, it is easy to see that the sequence $\{y_1 \otimes \otimes \ldots \otimes y_n\}$ converges to a limit y in H; we write $y = y_1 \otimes y_2 \otimes \ldots$. The map $y_1 \otimes \ldots \otimes y_n \to y_1 \otimes \ldots \otimes y_n \otimes x_{n+1} \otimes x_{n+2} \otimes \ldots$ defines a natural isometry of $H_1 \otimes \ldots \otimes H_n$ into H; in the same simple way a natural isomorphism

$$H_1 \otimes \ldots \otimes H_k \otimes \otimes_{n=k+1}^{\infty, \{x_n\}} H_n \leftrightarrow \otimes_{n=1}^{\infty, \{x_n\}} H_n$$

may be defined. If T_n is an operator in H_n, then

$$\overline{T}_n(y_1 \otimes y_2 \otimes \ldots \otimes y_n \otimes \ldots) = y_1 \otimes y_2 \otimes \ldots \otimes T_n y_n \otimes \ldots$$

defines an operator in H, and the map $T_n \to \overline{T}_n$ is easily seen to preserve norms, sums, products, and adjoints. If $\{A_n\}$ is a sequence of factors, A_n acting in the space H_n, then the W*-algebra A generated by the operators $\overline{T}_n$, $T_n \in A_n$, may be seen to be a factor, and is designated by the symbol

$$\otimes_{n=1}^{\infty, \{x_n\}} A_n.$$

It is readily seen that A is the weak closure of the class of operators of the form $T \otimes I$, where $T \in A_1 \otimes \ldots \otimes A_k$ and I is the identity operator in $\otimes_{n>k+1}^{\infty, \{x_n\}} H_n$.

If (S_1, μ_1) and (S_2, μ_2) are two measure spaces of total measure 1, and G_1 G_2 groups of measurability-preserving transformations on S_1, S_2, respectively, both satisfying the hypotheses of Lemmas 1 and 2, it is easily seen that

$$(4)\quad A_1(G_1, S_1, \mu_1) \otimes A_1(G_2, S_2, \mu_2) = A_1(G_1 \times G_2, S_1 \times S_2, \mu_1 \times \mu_2).$$

Indeed, $A_1(G_i, S_i, \mu_i)$, $i = 1, 2$, acts in the Hilbert space of functions on $S_i \times G_i$, where each point in G_i is assigned unit measure. The mapping

$$[f_1(s_1, g_1), f_1(s_2, g_2)] \to f_1(s_1, g_1) f_1(s_2, g_2)$$

represents $\mathcal{L}_2(S_1 \times S_2, G_1 \times G_2)$ as the Kronecker product $\mathcal{L}_2(S_1 \times G_1) \otimes \mathcal{L}_2(S_2 \times \times G_2)$, and realizes the algebra isomorphism given by equation (4). Now if we let $\delta(s_2, g_2) \equiv 1$ when $g_2 = e$, $\delta(s_2, g_2) \equiv 0$ when $g_2 \neq e$, then the mapping $f_1(s_1, g_1) \to \to f_1(s_1, g_1)\delta(s_2, g_2)$ becomes an isometric imbedding of $\mathcal{L}_2(S_1 \times G_1)$ into $\mathcal{L}_2(S_2 \times \times S_2, G_1 \times G_2)$; this isometry brings the algebra $A_1(G_1, S_1, \mu_1)$ in the first space into coincidence with the algebra $A_1(G_1, S_1, \mu_1) \otimes I$ in the second space, regarded as a Kronecker product.

Using these observations and the above remarks concerning infinite direct products, it is not hard to establish the following result.

LEMMA 4. Let $\{(S_i, \mu_i)\}$, $i \geq 1$, be a sequence of measure spaces of total measure 1, and $\{G_i\}$, $i \geq 1$, groups of measurability-preserving transformations, G_i acting on S_i. Let

$$S = S_1 \times S_2 \times \ldots$$

$$\mu = \mu_1 \times \mu_2 \times \ldots$$

$$G = G_1 \oplus G_2 \oplus \ldots$$

be the indicated direct products, where more precisely G is the countable direct limit of the groups $G_1 \oplus \ldots \oplus G_n$. Then

$$A_1(G, S, \mu) = \otimes_{n=1}^{\infty, \{\delta_n\}} A_1(G_n, S_n, \mu_n).$$

Here, $\delta_n(s_n, g_n) \equiv 1$ if $g_n = e$, $\delta_n(s_n, g_n) \equiv 0$ if $g_n \neq e$.

Sketch of proof. Write $G_1 \oplus \ldots \oplus G_n = G^{(n)}$, $S_1 \times \ldots \times S_n = S^{(n)}$, $\mu_1 \times \ldots \times \mu_n = \mu^{(n)}$. Then the mapping $j_n: f \mapsto f \otimes \delta_{n+1}$ is readily seen to imbed $\mathcal{L}_2(G^{(n)}, \mathcal{L}_2(S^{(n)}, \mu^{(n)}))$ into $\mathcal{L}_2(G^{(n+1)}, \mathcal{L}_2(S^{(n+1)}, \mu^{(n+1)}))$ in the following way: Identifying each function g of $[s_1, s_2, \ldots, s_n] \in S^{(n)}$ with the same function regarded as a function of $n+1$ variables, g being independent of the last variable, we obtain an imbedding $j_n: \mathcal{L}_2(S^{(n)}, \mu^{(n)}) \to \mathcal{L}_2(S^{(n+1)}, \mu^{(n+1)})$. Given a function $f \in \mathcal{L}_2(G^{(n)}, \mathcal{L}_2(S^{(n)}, \mu^{(n)}))$, we may then extend the function $j_n f$ from the group $G^{(n)}$ on which it is defined to the larger group $G^{(n+1)} \supset G^{(n)}$ simply by writing $(j_n f)(x) = 0$ if $x \notin G^{(n)}$. It is easy to see that this extended function is precisely $j_n f$.

A function $g(s_1, \ldots, s_n) \in \mathcal{L}_2(S^{(n)}, \mu^{(n)})$ may also be regarded as an element of $\mathcal{L}_2(S, \mu)$ which is independent of all but the first n coordinates of $[s_1, s_2, \ldots] \in S$, giving us an imbedding of $\mathcal{L}_2(S^{(n)}, \mu^{(n)})$ in $\mathcal{L}_2(S, \mu)$. Since the diagram

$$\begin{array}{ccc} \mathcal{L}_2(S^{(n)}, \mu^{(n)}) & \longrightarrow & \mathcal{L}_2(S^{(n+1)}, \mu^{(n+1)}) \\ & \searrow & \downarrow \\ & & \mathcal{L}_2(S, \mu) \end{array}$$

is commutative for all n, and since the images in $\mathcal{L}_2(S, \mu)$ of the various spaces $\mathcal{L}_2(S^{(n)}, \mu^{(n)})$ span $\mathcal{L}_2(S, \mu)$, it follows that $\mathcal{L}_2(S, \mu)$ may be identified with the direct limit space of the spaces $\mathcal{L}_2(S^{(n)}, \mu^{(n)})$ via the sequence $\{j_n\}$ of imbeddings. Since the countable group G is the union of the increasing sequence of subgroups G_n, it may be deduced from this that $\mathcal{L}_2(G, \mathcal{L}_2(S, \mu))$ is the direct limit of the spaces $\mathcal{L}_2(G^{(n)}, \mathcal{L}_2(S^{(n)}, \mu^{(n)}))$ via the sequence of imbeddings j_n. Thus we identify $\mathcal{L}_2(G, \mathcal{L}_2(S, \mu))$ with the direct product Hilbert space $\otimes_{n=1}^{\infty, \{\delta_n\}} \mathcal{L}_2(G_n, \mathcal{L}_2(S_n, \mu_n))$. This identification maps the algebra M of multiplications by functions of S onto the weak closure of the algebra of multiplications by functions of a finite number of coordinates

$s_1, s_2, \ldots, s_n$ of $s \in S$ and so identifies M as the Kronecker product $\otimes_{n=1}^{\infty, \{\delta_n\}} M_n$, where M_n is the algebra in $\mathcal{L}_2(G_n, \mathcal{L}_2(S_n, \mu_n))$ generated by multiplications by bounded functions on S_n. From these facts, the present lemma follows with relative ease if the definition of the generalized group algebra $A_1(G, S, \mu)$ (cf. (1) above) is used. See also Bures [2] for additional details. Q. E. D.

From Lemma 4 we perceive at once that the type III factor $R^{(x)}$ described above may be written as an infinite direct product

$$R^{(x)} = \otimes_{n=1}^{\infty, \{\delta\}} \rho^{(x)}$$

where $\rho^{(x)} = A_1(\gamma, \sigma, v_x)$. That is, $R^{(x)}$ is the product of infinitely many copies of $\rho^{(x)}$. As noted previously, $\rho^{(x)}$ is a 4-dimensional factor acting in a 4-dimensional space H, and is thus structurally isomorphic to $B(H_1) \otimes I$ acting in $H = H_1 \otimes H_2$, where both H_1 and H_2 are 2-dimensional. The vector belongs to H and, in order to perfect our account of $R^{(x)}$ as an infinite Kronecker product, we need to know the relationship of δ to the two factor spaces H_1 and H_2. This may be derived as follows. Let X_1 and X_2 be two n-dimensional Hilbert spaces, and let x be a vector in $X_1 \otimes X_2$. Then the expression $\varphi(T) = ((T \otimes I)x, x)$ defines a linear functional on $B(X_1)$, and $\varphi(T)$ is clearly nonnegative if T is hermitian and $T \geqslant 0$. It follows that there exists a nonnegative hermitian $P \in B(X_1)$ such that $\varphi(T) = \mathrm{tr}(TP)$. Diagonalizing P, we may conclude at once that there exist eigenvalues $\lambda_1 \geqslant \lambda_2 \geqslant \ldots \geqslant \lambda_n \geqslant 0$ and orthonormal basis vectors $\psi_1, \psi_2, \ldots, \psi_n$ of X_1 such that

$$((T \otimes I)x, x) = \sum_{i=1}^{n} \lambda_i (T\psi_i, \psi_i). \tag{5}$$

Using these basis vectors, we may find vectors: $\hat{\eta}_1, \ldots, \hat{\eta}_n \in X_2$ such that $x = \Sigma\, \psi_i \otimes \hat{\eta}_i$. Substituting this expression in (5), and replacing T by a dyadic operator $\psi_e \circ \hat{\eta}$, we find that $(\hat{\eta}_k, \hat{\eta}_e) = \lambda_k \delta_{ke}$; thus there exists an orthonormal basis $\eta_1, \ldots, \eta_n$ of X_2 such that $x = \sum_{i=1}^{n} \lambda_i^{1/2} (\psi_i \circ \eta_i)$. This makes it plain that the eigenvalues $\lambda_1, \ldots, \lambda_n$ are the sole invariants of position of a vector $x \in X = X_1 \otimes X_2$ relative to the Kronecker product decomposition of X. In the case in which we are given an n^2-dimensional factor A acting in an n^2-dimensional space X, these invariants may be calculated as follows: express the linear form (Lx, x), $L \in A$, as $\mathrm{tr}(LP)$ where $P \in A$ and tr is the unique invariant trace on A. Then $\lambda_1, \ldots, \lambda_n$ are the eigenvalues of the operator P. This recipe applies readily to the case in which $A = A_1(G, S, \mu)$ and in which S is finite and consists of n points $p_1, \ldots, p_n$, $\mu(p_j) = \mu_j$, and G is a group of n elements acting transitively on S. In this case, the normalized invariant trace on A is given by the formula

$$\mathrm{tr}(T) = \int_S m(s)\nu(ds), \quad T \in A_1(G, S, \mu), \tag{6}$$

where T is given by (1), $\tau(x) = b(x)U(x)$, $b(e)$ is the operation of multiplication by

$m(s)$, and ν is the unique G-invariant measure on the finite space S. From the definition of the function δ it is clear that

$$(T\delta, \delta) = \int_S m(s)\mu(ds), \tag{7}$$

where T and m are as above. Thus, if we let ρ be the Radon-Nikodym derivative of μ with respect to ν, so that $\rho(p_j) = \mu_j$, and P the corresponding multiplication operator, it follows on comparing (6) and (7) that $(T\delta, \delta) = \mathrm{tr}(TP)$. Hence the eigenvalues λ_i of (5) are in this case identical with the measures $\mu(p_j)$.

In the particular case of the 4-dimensional ring $p^{(x)}$ and its associated vector δ, it follows from what has just been said that there exists an isomorphism of $p^{(x)}$ onto $B(H_1) \otimes I$ mapping δ onto the vector $a_x = \lambda^{1/2}\varphi_1 \otimes \eta_1 + (1 - \lambda)^{1/2}\varphi_2 \otimes \eta_2$, where φ_1, φ_2 are orthonormal, η_1, η_2 are orthonormal, and $\lambda/(1 - \lambda) = x$. This observation allows us to represent the infinite product ring $R^{(x)}$ in an alternate form as

$$R^{(x)} = \otimes_{n=1}^{\infty, \{a_x\}} (B(H_1) \otimes I). \tag{8}$$

We may now, following Araki and Woods, proceed with the proof of Powers Theorem 3. If $T \in B(H_1) \otimes B(H_2)$, we let $T^{(n)}$ be the operator in $R^{(x)}$ defined by

$$T^{(n)}(y_1 \otimes y_2 \otimes \ldots) = y_1 \otimes y_2 \otimes \ldots \otimes Ty_n \otimes y_{n+1} \otimes \ldots .$$

Let J and K be the operators in $B(H_1)$ and $B(H_2)$ respectively defined by $J\varphi_1 = \varphi_2$, $J\varphi_2 = 0$ and $K\eta_2 = \eta_1$, $K\eta_1 = 0$. Then plainly $((J \otimes I) - x^{1/2}(I \otimes K))a_x = 0$, and $(x^{1/2}(J^* \otimes I) - (I \otimes K^*))a_x = 0$. Thus, putting $J_n = (J \otimes I)^{(n)}$, $K_n = (I \otimes K)^{(n)}$, we obtain uniformly bounded sequences of operators $J_n \in R^{(x)}$, $K_n \in (R^{(x)})'$, i.e. in the commutator ring of $R^{(x)}$. Moreover, since vectors of the form $y_1 \otimes \ldots \otimes y_n \otimes a_x \otimes a_x \otimes a_x \otimes \ldots$ span the space H in which $R^{(x)}$ acts, the following limiting relations hold for each vector $v \in H$:

$$|J_n v| \to \lambda^{1/2}|v|; \quad |K_n v| \to (1 - \lambda)^{1/2}|v|;$$

$$|J_n^* v| \to (1 - \lambda)^{1/2}|v|; \quad |K_n^* v| \to \lambda^{1/2}|v|; \tag{9}$$

$$|(J_n - xK_n)v| \to 0; \quad |(xJ_n^* - K_n^*)v| \to 0.$$

Suppose now that $R^{(y)}$ is isomorphic to $R^{(x)}$. Then a corresponding sequence of operators may be associated with $R^{(y)}$. By definition, the infinite Kronecker product ring $R^{(y)}$ is the weak closure of the sequence of subrings $(\otimes_{n=1}^{m} \rho^{(y)}) \otimes I = R_m$, I being the identity of the ring $\otimes_{n=m+1}^{\infty} \rho^{(y)}$. The Kaplansky density theorem (cf. [18], Chapter I, section 5, Theorem 3, p. 46) therefore tells us that each uniformly bounded sequence of operators in $R^{(y)}$ can be approximated by a uniformly bounded sequence whose elements belong to $\bigcup_{m=1}^{\infty} R_m$. Similarly, each uniformly bounded sequence of operators in the commutator ring $(R^{(y)})'$ can be approximated by a uniformly bounded sequence whose elements belong to $\bigcup_{m=1}^{\infty}(\otimes_{n=1}^{m}(\rho^{(y)})' \otimes I)$. Thus, considering (9), and taking v to be the vector $a_y \otimes a_y \otimes \ldots$, it would follow that for each positive ϵ there existed integers m and operators $E \in \rho_m^{(y)} = \otimes_{n=1}^{m} \rho^{(y)}$, $F \in (\rho_m^{(y)})' = \otimes_{n=1}^{m} (\rho^{(y)})'$ such that the vector $a_y \otimes \ldots \otimes a_y = \beta$ satisfies

$$|E\beta| > 1;$$

$$(10) \qquad |(E - x^{1/2}F)\beta| < \epsilon;$$

$$|(x^{1/2}E^* - F^*)\beta| < \epsilon.$$

The factor $\rho_m^{(y)}$ is 4^m-dimensional and acts in a 4^m-dimensional space H, so that we may map H onto a Kronecker product $H_1 \otimes H_2$ in such a way as to represent $\rho_m^{(y)}$ as $B(H_1) \otimes I$ and its commutator $(\rho_m^{(y)})'$ as $I \otimes B(H_2)$. In this representation, the vector β will appear according to our earlier analysis as

$$(11) \qquad \beta = \sum_{i=1}^{M} \hat{\lambda}_i^{1/2} \varphi_i \otimes \hat{\eta}_i,$$

where $M = 2^m$, and where $\{\varphi_i\}$ and $\{\eta_i\}$ are orthonormal. Moreover, E and F will respectively appear as

$$(12) \qquad E = e \otimes I, \quad F = I \otimes f.$$

The coefficients $\hat{\lambda}_i$ may readily be calculated from the corresponding expression $a_y = v^{1/2} \varphi_1 \otimes \tilde{\eta}_1 + (1 - v)^{1/2} \tilde{\varphi}_2 \otimes \tilde{\eta}_2$ for the vector a_y, and we see in this way that each $\hat{\lambda}_i$ has the form $v^j(1 - v)^{m-j}$; this particular eigenvalue being repeated $\binom{m}{j}$ times. Note the fact, to be used below in an essential manner, that $v/(1 - v) = y$. Thus each ratio between a pair of the eigenvalues $\hat{\lambda}_i$ is a power $y^{\pm j}$ of y.

Next, note the following relationships which follow from (11) and (12).

$$(13) \qquad \epsilon^2 > |(E - xF)\beta|^2 = \sum_{i,j=1}^{M} |((E - xF)\beta, \hat{\varphi}_1 \otimes \hat{\eta}_j)|^2$$

$$= \sum_{i,j=1}^{M} |\hat{\lambda}_j^{1/2}(e\hat{\varphi}_j, \hat{\varphi}_i) - x^{1/2}\lambda_i^{1/2}(f\hat{\eta}_i, \hat{\eta}_j)|^2.$$

Calculating similarly for the adjoint operators we find

$$\epsilon^2 > \sum_{i,j=1}^{M} |x^{1/2}\hat{\lambda}_j^{1/2}(e\hat{\varphi}_j, \hat{\varphi}_i) - \hat{\lambda}_i^{1/2}(f^*\hat{\eta}_i, \hat{\eta}_j)|^2,$$

i.e.

$$(14) \qquad \epsilon^2 > \sum_{i,j=1}^{M} |x^{1/2}\hat{\lambda}_i^{1/2}(e\hat{\varphi}_j, \hat{\varphi}_i) - \hat{\lambda}_j^{1/2}(f\hat{\eta}_i, \hat{\eta}_j)|^2.$$

Combining (13) and (14) by use of the triangle inequality we find

$$(2\epsilon)^2 > \sum_{i,j=1}^{M} (\hat{\lambda}_i^{1/2} + x^{1/2}\hat{\lambda}^{1/2})^2 \, |(e\hat{\varphi}_j, \hat{\varphi}_i) - (f\hat{\eta}_i, \hat{\eta}_j)|^2,$$

and, since $\hat{\lambda}_i$ and x are positive, that

$$(2\epsilon)^2 > \sum_{i,j=1}^{M} (x^{1/2}\hat{\lambda}_i^{1/2})^2 \, |(e\hat{\varphi}_j, \hat{\varphi}_i) - (f\hat{\eta}_i, \hat{\eta}_j)|^2. \tag{15}$$

Combining (15) and (13) by use of the triangle inequality we obtain

$$(3\epsilon)^2 > \sum_{i,j=1}^{M} |\hat{\lambda}_j^{1/2} - x^{1/2}\hat{\lambda}_i^{1/2}|^2 (e\hat{\varphi}_j, \hat{\varphi}_i)^2.$$

On the other hand, from the first inequality in (9) we have

$$1 \leqslant \sum_{i,j=1}^{n} (E\beta, \hat{\varphi}_i \otimes \hat{\eta}_j)^2 = \sum_{i,j=1}^{\infty} \eta_j (e\hat{\varphi}_i, \hat{\varphi}_j)^2.$$

Put $c_{ij} = \hat{\lambda}_j |(e\hat{\varphi}_i, \hat{\varphi}_j)|^2$; then $\sum_{i,j=1}^{M} c_{ij} \geqslant 1$, and

$$\sum_{i,j=1}^{n} |1 - x^{1/2}\hat{\lambda}_i^{1/2}/\hat{\lambda}_j^{1/2}|^2 c_{ij} < (3\epsilon).$$

It follows that some coefficient $|1 - x^{1/2}\hat{\lambda}_i^{1/2}/\hat{\lambda}_j^{1/2}|$ is less than 3ϵ; therefore some ratio $\hat{\lambda}_i^{1/2}/\hat{\lambda}_j^{1/2}$ is closely approximated by x^{-1}. But, as we have already seen, each such ratio has the form $y^{\pm n}$. Letting $\epsilon \to 0$, and observing that $0 < x < 1$ and $0 < y < 1$, it follows that we must have $x = y^n$ for some exponent $n \geqslant 1$. But by symmetry, we must also have $y = x^k$ for some exponent $k \geqslant 1$. Thus the equality $x = y$ follows from the isomorphism of $R^{(x)}$ and $R^{(y)}$, completing the proof of Powers Theorem 3. Q. E. D.

Powers and Araki-Woods have investigated a wider class of type III factors definable by infinite Kronecker products of finite-dimensional factors and obtained other interesting results. Let $\Lambda = \{\lambda_1, \ldots, \lambda_n\}$ denote any set of n nonnegative numbers whose sum is 1, so that $\lambda_1, \ldots, \lambda_n$ serve to define a measure $\mu(\Lambda)$ on a space $S(\Lambda)$ of n points $p_1, \ldots, p_n$. We call Λ a positive normalized vector. As a matter of convenience we will suppose throughout that Λ has an *even* number of components. Let $G(\Lambda)$ be the group of transformations of $S(\Lambda)$ generated by the circular shift $p_1 \to \to p_2 \to \ldots \to p_{n-1} \to p_n \to p_1$. Let $\Lambda_1, \Lambda_2, \ldots$ be any sequence of vectors Λ. Then we may consider the

space $S = S(\Lambda_1) \times S(\Lambda_2) \times \ldots$

measure $\mu = \mu_1 \times \mu_2 \times \ldots$

group $G = G(\Lambda_1) \times G(\Lambda_2) \times \ldots$

and the factor $A_1(G, S, \mu)$, whose derivation we emphasize by denoting it as $R(\Lambda_1, \Lambda_2, \Lambda_3, \ldots)$. These factors are the objects of the Powers and Araki-Woods investigations. Concerning them we may note the following elementary facts.

LEMMA 5. We have:

(a) If $\{\hat{\Lambda}_i\}$ is any permutation of $\{\Lambda_i\}$, then $R(\hat{\Lambda}_1, \hat{\Lambda}_2, \ldots)$ and $R(\Lambda_1, \Lambda_2, \ldots)$ are isomorphic.

(b) If $\{\tilde{\Lambda}_i\}$ is a second sequence of vectors Λ, then $R(\tilde{\Lambda}_1, \tilde{\Lambda}_2, \ldots) \otimes R(\Lambda_1, \Lambda_2, \ldots)$ is isomorphic to $R(\tilde{\Lambda}_1, \Lambda_1, \tilde{\Lambda}_2, \Lambda_2, \ldots)$.

(c) Let $\Lambda_1 \circ \Lambda_2 \circ \ldots \circ \Lambda_n$ denote the set of all products of n elements, one chosen from each of Λ_j. Then $R(\Lambda_1, \Lambda_2, \ldots)$ and $R(\Lambda_1 \circ \Lambda_2 \circ \ldots \circ \Lambda_{n_i}, \Lambda_{n_{i+1}} \circ \Lambda_{n_{i+2}} \circ \ldots \circ \Lambda_{n_2}, \ldots)$ are isomorphic.

Note that $R^{(x)} = R(\Lambda^{(x)}, \Lambda^{(x)}, \ldots)$, where $\Lambda^{(x)} = (x/(1+x), 1+x)$. Thus, the following statement results at once from (b) of the preceding Lemma.

COROLLARY 6. $R^{(x)} \otimes R^{(x)}$ is isomorphic to $R^{(x)}$.

The argument used to establish the representation (8) for the factor $R^{(x)} = R(\Lambda^{(x)}, \Lambda^{(x)}, \ldots)$ applies with minimal change to the more general factors $R(\Lambda_1, \Lambda_2, \ldots)$, and leads to the following result.

LEMMA 7. (Bures) For each normalized vector $\Lambda = \{\lambda_1, \ldots, \lambda_n\}$ let $H_1^{\wedge}$ and $H_2^{\wedge}$ be two vector spaces of the same dimension n as Λ, and let $\varphi_1, \ldots, \varphi_n$ and $\eta_1, \ldots, \eta_n$ be orthonormal bases of H_1 and H_2 respectively. Further let $a^{\Lambda n} = \sum_{k=1}^{n} \lambda_k^{1/2} \varphi_k \otimes \eta_k$. Let $\Lambda_1, \Lambda_2, \ldots$ be a sequence of normalized vectors, and $a_n = a^{\Lambda n}$. $R(\Lambda_1, \Lambda_2, \ldots)$ is isomorphic to the Kronecker product

$$\bigotimes_{n=1}^{\infty, \{a_n\}} (B(H_1{}^{\Lambda n}) \otimes I).$$

We also note for use below that the argument used to prove Theorem 3 has a consequence for the rings $R(\Lambda_1, \Lambda_2, \ldots)$ somewhat more extensive than that explicitly stated. The existence of the operators E and F of (10) above follows not only from the hypothesis $R^{(y)} \sim R^{(x)}$, but even from the weaker hypothesis $R^{(y)} \sim A \otimes R^{(x)}$, A being any factor. Once these operators are known to exist, we can follow the remainder of the proof of Theorem 3 with no change, to conclude that the hypothesis $R^{(y)} \sim A \otimes R^{(x)}$ implies that x has the form $y^{\pm n}$.

Other interesting Kronecker product relations can be deduced from the following Lemma, which follows rather easily from Lemma 7.

LEMMA 8. Let $\{\hat{\Lambda}_j\}$ and $\{\Lambda_j\}$ be two sequences of positive normalized vectors. Suppose that $\hat{\Lambda}_j$ and Λ_j are of the same dimension for $j \geq 1$, and that $\sum_{j=1}^{\infty} |\hat{\Lambda}_j - \Lambda_j| < \infty$ where we write $|v|$ for the ℓ_2-norm of v. Then $R(\hat{\Lambda}_1, \hat{\Lambda}_2, \ldots)$ and $R(\Lambda_1, \Lambda_2, \ldots)$ are isomorphic.

Proof. Let all notations be as in Lemma 7, and put $\beta_n = a^{\hat{\Lambda}_n}$. Then $(\beta_n, a_n) = (\hat{\Lambda}_n^{1/2}, \Lambda_n^{1/2}) > 0$, so that $1 - (\beta_n, a_n) = (|\beta_n|^2 + |a_n|^2 - 2(\beta_n, a_n))/2 = |\beta_n - a_n|^2/2$. Since $(\hat{\lambda}_i^{1/2} - \lambda_i^{1/2})^2 \leq |\hat{\lambda}_i^{1/2} - \lambda_i^{1/2}|(\hat{\lambda}_i^{1/2} + \lambda_i^{1/2}) = |\hat{\lambda}_i - \lambda_i|$, our hypotheses imply that $\sum_{n=1}^{\infty} |1 - (\beta_n, a_n)|$ converges. Thus, as observed in our general discussion concerning infinite Kronecker products, the sequence of vectors $\{\beta_1 \otimes \ldots \otimes \beta_n \otimes a_{n+1} \otimes a_{n+2} \otimes \ldots\}$ converges to a limit $\beta_1 \otimes \beta_2 \otimes \ldots$ in $\otimes_{n=1}^{\infty, \{a_k\}} H_n$, where $H_n = H_1{}^{\Lambda_n} \otimes H_2{}^{\Lambda_n}$ in the notation of Lemma 7. The map which sends each vector $x_1 \otimes \ldots \otimes x_n \otimes \beta_{n+1} \otimes \beta_{n+2} \otimes \ldots$ of the space $\otimes_{n=1}^{\infty, \{\beta_n\}} H_n$ into the vector of $\otimes_{n=1}^{\infty, \{a_k\}} H_n$ having exactly the same formula then is an isometry of $\otimes_{n=1}^{\infty, \{\beta_n\}} H_n$ onto $\otimes_{n=1}^{\infty, \{a_n\}} H_n$. Plainly an isomorphism of this special form puts the algebra $\otimes_{n=1}^{\infty, \{a_n\}} (B(H_1{}^{\Lambda_n}) \otimes I)$ generated by operators

$$x_1 \otimes \ldots \otimes x_n \otimes \beta_{n+1} \otimes \ldots \to x_1 \otimes \ldots \otimes (T \otimes I)x_n \otimes \beta_{n+1} \otimes \ldots$$

into unitary correspondence with the algebra $\otimes_{n=1}^{\infty, \{\beta_n\}} (B(H_1{}^{\Lambda_n}) \otimes I)$, establishing the isomorphism asserted in our lemma. Q. E. D.

Using Lemma 8 as a basic tool, Araki and Woods give the following definition and theorem.

DEFINITION 9. Let A be a W*-algebra. Then $r_\infty(A)$, the *asymptotic ratio set* for A, is the set of $x \in [0, \infty)$ such that A and $A \otimes R^{(x)}$ are isomorphic

THEOREM 10. Let A be a W*-algebra. Then:

(i) The set of finite nonzero elements of $r_\infty(A)$ is a multiplicative group.

(ii) If A has the form $R(\Lambda_1, \Lambda_2, \ldots)$, then $r_\infty(A)$ is a closed set.

(iii) $r_\infty(R^{(x)}) = \{x^{\pm n}: n = 0, 1, 2, \ldots\} \cup \{0\}$.

It follows as a corollary of Theorem 10 that each A of the form $R(\Lambda_1, \Lambda_2, \ldots)$ has an asymptotic ratio set of one of the following forms.

i. $r_\infty(A)$ is null;

ii. $r_\infty(A) = \{1\}$;

iii. $r_\infty(A) = \{0, 1\}$;

iv. $r_\infty(A) = [0, \infty)$;

v. $r_\infty(A) = \{x^{\pm n}: n = 0, 1, 2, \ldots\} \cup \{0\}$.

Cases (i) and (ii) describe type II factors, and examples of both cases may be given. Araki and Woods show that uncountably many factors A belonging to case (iii) exist; these factors are distinguishable by the invariant $P(A) = \{x: R^{(x)} \sim R^{(x)} \otimes A\}$. The factors $A = R(\Lambda_1, \Lambda_2, \ldots)$ belonging to case (iv) are all shown to be isomorphic; a typical factor of this class is $R^{(x)} \otimes R^{(y)}$, with log x and log y irrational multiples of each other. Similarly, the factors $A = R(\Lambda_1, \Lambda_2, \ldots)$ belonging to case (v) are isomorphic to $R^{(x)}$.

We refer the reader to Araki-Woods [4] for proofs of these last results. In this paper we shall confine ourselves to giving the proof of Theorem 10. In preparation, we state the following definition and lemma.

DEFINITION 11. Let $L = [\Lambda_1, \Lambda_2, \ldots]$ be a sequence of positive normalized vectors and let $\Lambda \circ \Lambda'$ denote the multiplication operator of Lemma 5(c). Let $\Lambda^x = [x(1+x)^{-1}, (1+x)^{-1}]$. For each set $S = \{i_1, \ldots, i_n\}$ of integers, let $\Lambda_S = \Lambda_{i_1} \circ \Lambda_{i_2} \circ \ldots \circ \Lambda_{i_n}$. Call x an *asymptotic ratio* for L if there exists a sequence $\{S_j\}$ of disjoint sets of integers and a sequence $\{\hat{\Lambda}_n\}$ of positive normalized vectors such that Λ_{S_n} and $\Lambda^x \circ \hat{\Lambda}_n$ have the same dimension and

$$|\Lambda_{S_n} - \Lambda^x \circ \hat{\Lambda}_n| \to 0,$$

where the indicated vector norm is the sum of absolute values of components.

LEMMA 12. If in Definition 11 x is an asymptotic ratio for L, then $R(\Lambda_1, \Lambda_2, \ldots) \sim R(\Lambda_1, \Lambda_2, \ldots) \otimes R^{(x)}$.

Proof. Passing if necessary to a subsequence, we may suppose that $\sum_{n=1}^{\infty} |\Lambda_{S_n} - (\Lambda^x \circ \Lambda_n)^{1/2}| < \infty$. If $\bar{L}$ denotes the complement in L of the sequence $\{\Lambda_{S_n}\}$ then it follows using Lemma 5 (a), (b), and (c) that

$$R(L) \sim R(\bar{L}) \otimes R(\Lambda_{S_1}, \Lambda_{S_2}, \ldots).$$

But the isomorphism

$$R(\Lambda_{S_1}, \Lambda_{S_2}, \ldots) \sim R(\Lambda^x \circ \hat{\Lambda}_1, \Lambda^x \circ \hat{\Lambda}_2, \ldots)$$

follows from Lemma 8, while

$$R(\Lambda^x \circ \hat{\Lambda}_1, \Lambda^x \circ \hat{\Lambda}_2, \ldots) = R(\hat{\Lambda}_1, \hat{\Lambda}_2, \ldots) \otimes R^{(x)}$$

follows from Lemma 5. Thus

$$R(L) \sim R(\overline{L}) \otimes R(\hat{\Lambda}_1, \hat{\Lambda}_2, \ldots) \otimes R^{(x)}$$

and since we have already seen that $R^{(x)} \sim R^{(x)} \otimes R^{(x)}$, we find that $R(L) \sim R(L) \otimes$ $\otimes R^{(x)}$, completing the proof. Q. E. D.

Our next aim is to show that the asymptotic ratio relationship between a number x and a sequence L described by Definition 11 is implied by the formally weaker relationship set forth in the following definition.

DEFINITION 13. Let x and L be as in Definition 11. Then x is called a *substantial ratio* for L if there exists a sequence of disjoint sets S_n of integers having the following properties:

i. Λ_{S_n} (cf. Definition 11 for notation) contains two subsets A_n and B_n such that $0 \notin A_n$ and such that $\sum_{\lambda \in A_n} \lambda$ is bounded below.

ii. There exists a 1-1 map $\varphi_n: A_n \to B_n$ such that $\max\{|x - \varphi_n\lambda/\lambda|: \lambda \in A_n\} \to$ $\to 0$.

LEMMA 14. If x is a substantial ratio for L, then x is an asymptotic ratio for L.

Before giving the proof of Lemma 14, we develop various of its consequences, one of which is Theorem 10.

COROLLARY 15. If the positive normalized vector Λ has two components whose ratio is x, then $R(\Lambda, \Lambda, \ldots)$ has x as an asymptotic ratio.

Proof. Let the two components be λ, μ, and put $A_n = \{\lambda\}$, $B_n = \{\mu\}$ for all n in Definition 13 and Lemma 14. Q. E. D.

COROLLARY 16. $R^{(x)} \otimes R^{(y)}$ has all the quantities $x^{\pm n}y^{\pm m}$ as asymptotic ratios.

Proof. Using Lemma 5 we obtain the isomorphism

$$R^{(x)} \otimes R^{(y)} \sim R((\Lambda^x)^n \circ (\Lambda^y)^m, (\Lambda^x)^n \circ (\Lambda^y)^m, \ldots).$$

Since $(\Lambda^x)^n \circ (\Lambda^y)^m$ has components whose ratio is $x^{\pm n}y^{\pm m}$ the present corollary follows from Corollary 15. Q. E. D.

COROLLARY 17. The nonzero asymptotic ratios of a W*-algebra A form a multiplicative group.

Proof. If $A \sim A \otimes R^{(x)}$ and $A \sim A \otimes R^{(y)}$, then $A \sim A \otimes R^{(x)} \otimes R^{(y)}$. If $z = x^{\pm n}y^{\pm m}$, then by Corollary 16 $R^{(x)} \otimes R^{(y)} \sim R^{(x)} \otimes R^{(y)} \otimes R^{(z)}$, and thus $A \sim$ $\sim A \otimes R^{(z)}$. Q.E.D.

We now sketch the proof of Theorem 10. Conclusion (i) is merely Corollary 17. Since $R^{(x)} \sim R^{(x)} \otimes R^{(x)}$ is elementary, we have $x \in r_\infty(R^{(x)})$ and thus $x^{\pm n} \in$ $\in r_\infty(R^{(x)})$ by Corollary 17. Since $R^{(0)}$ is easily seen to be type I the factor $B(H_1) \otimes$ $\otimes I$ in the Kronecker product of two infinite dimensional Hilbert spaces, and since each type III factor is readily seen to be isomorphic to its own product with any type I factor, we have $0 \in r_\infty(R^{(x)})$ also.

The remark made immediately following Lemma 7 shows that if $R^{(y)} \sim R^{(y)} \otimes$ $\otimes R^{(x)}$, then $x = y^{\pm n}$. Thus if $0 \neq t \in r_\infty(R^{(x)})$, then $t \in x^{\pm n}$. It only remains to establish that the set $r_\infty(A)$ is closed, which is conclusion (ii) of Theorem 10. If any nonzero element x belongs to $r_\infty(A)$, then $A \sim A \otimes R^{(x)}$, and since $R^{(x)}$ is of type III it follows that A is of type III also. Thus, as we have seen, $0 \in r_\infty(A)$. This shows that to obtain (ii) one need only show that the collection of nonzero elements of $r_\infty(A)$ is closed. For the proof of this fact, which uses a more precise version of the argument used in the proof of Theorem 3, we refer the reader to Araki-Woods [4].

We conclude with the proof of Lemma 14.

Proof. Let all notations be as in Definitions 11 and 13. Let $\epsilon > 0$, and such that $\max \{|x - \varphi_n\lambda/\lambda|: \lambda \in A_n\} < \epsilon, j = 1, \ldots, n$. If A is any finite set of real numbers, write ΣA for the sum of the elements of A. Since (cf. Definition 13) ΣA_n is bounded below, we can choose disjoint $S_1, \ldots, S_n$ such that

$$\prod_{i=1}^{n} (1 - \Sigma A_i - \Sigma B_i) < \epsilon.$$

Write $S = S_1 \cup \ldots \cup S_n$, so that $\Lambda_S = \Lambda_{S_1} \circ \ldots \circ \Lambda_{S_n}$; then the components λ of Λ_S are products $\lambda_1 \cdot \lambda_2 \cdot \ldots \cdot \lambda_n$ of components of $\Lambda_{S_1}, \ldots, \Lambda_{S_n}$. Classify the components of Λ_S into three sets σ_-, σ_0, σ_+ as follows:

σ_0 contains all the elements $\lambda = \lambda_1 \cdot \lambda_2 \cdot \ldots \cdot \lambda_n$ of Λ_S which are such that no $\lambda_j \in A_j \cup B_j$.

σ_+ contains all the elements $\lambda = \lambda_1 \cdot \lambda_2 \cdot \ldots \cdot \lambda_n$ of Λ_S such that the first j belonging to $A_j \cup B_j$ belongs to A_j. For such λ, we put

$$\varphi\lambda = \lambda_1 \cdot \lambda_2 \cdot \ldots \cdot \lambda_{j-1} \cdot \varphi_j\lambda_j \cdot \lambda_{j+1} \cdot \ldots \cdot \lambda_n.$$

σ_- contains all the elements $\lambda = \lambda_1 \cdot \lambda_2 \cdot \ldots \cdot \lambda_n$ of Λ_S such that the first j belonging to $A_j \cup B_j$ belongs to B_j.

Note that φ is a 1-1 mapping of σ_+ onto σ_-, and that $\max \{|x - \varphi\lambda/\lambda|: \lambda \in \sigma_+\} <$ $< \epsilon$. Now

$$\Sigma\,\sigma_0 \leqslant \prod_{j=1}^{n}\ \Sigma\,(A_j \cup B_j)',$$

where X' denotes the complement of X. Thus we have

$$\Sigma\,\sigma_0 \leqslant \prod_{j=1}^{n}\ (1 - \Sigma\,A_j - \Sigma\,B_j) < \epsilon.$$

Since A_n and B_n are equally numerous, and since we assume that all the vectors Λ_n have an even number of components, σ_0 contains an even number of elements and may be divided into two parts σ_1 and σ_2. Now let the components of the vector M be an enumeration in decreasing order of all elements which belong either to σ_+ or to σ_- and let $\hat{\Lambda}$ be proportional to M but normalized so that $\Sigma\,\hat{\Lambda} = 1$. Let M_1 be the vector, of twice the dimension of M, which may be written as $M \otimes xM$, and let M_2 be proportional to M_1 but normalized. Since $|x\lambda - \varphi\lambda| < \epsilon\lambda$ for $\lambda \in \sigma_+$, and since $\Sigma\,\sigma_0 < \epsilon$, we have $|\Lambda_S - M_1| < \epsilon$. Since Λ_S is normalized, it follows that $|1 - \Sigma\,M_1| < \epsilon$, so that $|M_1 - (\Sigma\,M_1)^{-1}M_1| < 2\epsilon$, i.e., $|M_1 - M_2| < 2\epsilon$. Thus $|\Lambda_S - M_2| < 3\epsilon$. Since $\Lambda^x \circ \hat{\Lambda}$ and M_2 are clearly proportional to each other, and since both are normalized, they must be identical, so that $|\Lambda_S - \Lambda^x \circ \hat{\Lambda}| < 3\epsilon$, and Lemma 14 clearly follows from this. Q. E. D.

REFERENCES

1. J. Schwartz, W*-Algebras, Notes on Mathematics and its Applications, Gordon and Breach Publishers, New York, 1968.
2. D. J. C. Bures, Certain factors constructed as infinite tensor products, Compositio Math., v. 15 (1962), pp. 169-238.
3. R. T. Powers, Representations of uniformly hyperfinite algebras and their associated von Neumann rings, Ann. of Math. [2], v. 86 (1967), pp. 138-171.
4. H. Araki and E. J. Woods, A classification of factors, Publ. RIMS, Kyoto University Ser. A., v. 4 (1968), pp. 51-130.
5. W. Krieger, On constructing non-isomorphic factors of type III, J. Functional Analysis, to appear.
6. W. Krieger, On a class of hyperfinite factors that arise from null-recurrent Markov chains, Preprint, Oct. 1969.
7. L. Pukanszky, Some examples of factors, Publicationes Mathematicae, v. 4 (1956), pp. 135-156.
8. J. Schwartz, Non-isomorphism of a pair of factors of type III, Comm. Pure and Appl. Math., v. 16 (1963), pp. 111-120.
9. W. Ching, Non-isomorphic, non-hyperfinite factors, Canadian J. Math., to appear.
10. J. Dixmier and E. C. Lance, Deux nouveaux facteurs de type II, Inventiones Math., to appear.
11. S. Sakai, Asymptotically abelian II_1 factors, to appear.
12. G. Zeller-Meier, Deux autres facteurs de type II_1, Inventiones Math., to appear.
13. J. Schwartz, Two finite, non-hyperfinite, non-isomorphic factors, Comm. Pure and Appl. Math., v. 16 (1963), pp. 19-26.

14. D. McDuff, A countable infinity of II_1 factors, Preprint, 1968.
15. D. McDuff, Uncountably many II_1 factors, Preprint, 1968.
16. D. McDuff, Central sequences and the hyperfinite factor, Preprint, 1969.
17. D. McDuff, On residual sequences in a II_1 factor, Preprint, 1969.
18. J. Dixmier, Les Algebres d'Operateurs dan l'Espace Hilbertien, Gauthier-Villars, Paris, 1957.

AN UNCOUNTABLE NUMBER OF II_1, II_∞ FACTORS

Shôichirô Sakai[1] – *University of Pennsylvania*

§1. INTRODUCTION. Recently, great progress has been made in the investigation of the isomorphism classes of II_1 factors ([1], [2], [4], [6], [7], [8]). In particular, McDuff [4] proved the existence of a countably infinite number of II_1 factors on a separable Hilbert space.

In this paper, by using the method of McDuff, we shall show the existence of an uncountable number of non-isomorphic II_1 factors on a separable Hilbert space. Moreover, by using this result and tensor products, we shall show the existence of an uncountable number of II_∞ factors on a separable Hilbert space.

Concerning III factors, Powers [11] has shown the existence of an uncountable number.

Added in proof. After writing this paper, the author received two papers by McDuff [9], [10] in which she proves the existence of an uncountable number of II_1 factors. Her construction is different from ours.

§2. CONSTRUCTION OF EXAMPLES. Suppose $G_1, G_2, \ldots$ and $H_1, H_2, \ldots$ are two sequences of groups. We denote by $(G_1, G_2, \ldots; H_1, H_2, \ldots)$ the group generated by the G_i's and the H_i's with additional relations that H_i, H_j commute elementwise for $i \neq j$ and G_i, H_j commute elementwise for $i \leqslant j$. This situation was considered in [2].

Let $L_1 = (G_1, G_2, \ldots; H_1, H_2, \ldots)$ with $G_i = Z$ and $H_i = Z$, all i, where Z is the infinite cyclic group. Define L_k inductively: $L_k = (G_1, G_2, \ldots; H_1, H_2, \ldots)$, with $G_1 = Z$, $H_i = L_{k-1}$ for all i and $k \geqslant 2$.

Let I be the set of all positive integers and let I_1 be a sequence of positive integers. Let $M_n(I_1) = \sum_{i=1}^{n} \oplus L_{p_i}$, if $I_1 = (p_1, p_2, \ldots)$ is infinite, and $M_n(I_1) = \sum_{i=1}^{n} \oplus \oplus L_{p_i}$ for $n \leqslant n_0$ and $M_n(I_1) = M_{n_0}$ for $n > n_0$ if $I_1 = (p_1, p_2, \ldots, p_{n_0})$ is finite.

Let $G[I_1] = (G_1, G_2, \ldots; H_1, H_2, \ldots)$ with $G_i = Z$ and $H_i = M_i(I_1)$ for all i. For a discrete group G, U(G) is the W*-algebra generated by the left regular representation of G. We shall prove the following theorem.

THEOREM 1. Let $I_1 = (p_i)$ and $I_2 = (q_i)$ be two sequences of I such that I_2 contains a positive integer $q \notin I_1$. Then $U(G[I_1])$ is not *-isomorphic to $U(G[I_2])$. In particular, $U(G[I_1])$ is not *-isomorphic to $U(G[I_2])$ if I_1 and I_2 are distinct

[1] Supported in part by the National Science Foundation.

sets of positive integers.

As a corollary, we have

COROLLARY 1. There exists an uncountable number of non-isomorphic II_1 factors on a separable Hilbert space.

To prove Theorem 1, we shall provide some preliminary considerations.

DEFINITION 1 ([4]). For a W*-algebra $U(G)$ we shall write $(U(G))_1$ to denote the unit sphere of $U(G)$. If B and C are subalgebras of $U(G)$ and $\delta > 0$, then we shall write $B \overset{\delta}{\subset} C$ to mean that given any $T \in (B)_1$, there exists some $S \in (C)_1$ with $\| T - S \|_2 < \delta$, where $\| x \|_2$ is the $\mathcal{L}_2(G)$-norm of x when $U(G)$ is embedded into $\mathcal{L}_2(G)$ canonically.

Let $A = U(G)$. A bounded sequence (T_n) of elements of A is called a *central sequence* if for all $X \subset A$, $\| [X, T_n] \|_2 \to 0$ $(n \to \infty)$, where $[\ ,\ ]$ is the Lie product. Central sequences (T_n), (T_n') in A are called equivalent, if $\| T_n - T_n' \|_2 \to 0$ $(n \to \infty)$.

Let H be a subgroup of a group of G. Then H is called *strongly residual* in G if it satisfies the following conditions: there exist a subset S of the complement $G \backslash H$ of H and elements g_1, g_2 of G such that

(i) $g_1 H g_1^{-1} = H$;

(ii) $S \cup g_1 S g_1^{-1} = G \backslash H$;

(iii) $\{g_2^n S g_2^{-n}\}$, $n = 0, \pm 1, \pm 2, \ldots$, forms a family of disjoint subsets of $G \backslash H$.

From the above definition, we can easily see that only one strongly residual subgroup of a commutative group G is G itself; in this case, S is the empty set.

LEMMA 1 ([4]). Let G_i, $i = 1, 2, \ldots, n$ be a finite family of groups and let H_i, $i = 1, 2, \ldots, n$ be a subgroup of G_i. Suppose that H_i is strongly residual in G_i for each i; then $\sum_{i=1}^{n} \oplus H_i$ is strongly residual in $\sum_{i=1}^{n} \oplus G_i$.

Let H be a strongly residual subgroup of G; then H must contain the center of G. Let $\{H_n\}$ be a sequence of subgroups of G. $\{H_n\}$ is called a *residual sequence* of G if it satisfies the following conditions:

(i) H_n is strongly residual in G;

(ii) $H_n = H_{n+1} \oplus K_n$, where K_n is a subgroup of G;

(iii) $\bigcup_{n=1}^{\infty} H_n' = G$, where H_n' is the commutant of H_n in G.

Let G_i, $i = 1, 2, \ldots, m$, be a finite family of groups and let $\{H_{i,n}\}$, $i = 1, 2, \ldots, m$, be a residual sequence of G_i; then $\{\sum_{i=1}^{m} \oplus H_{i,n}\}$ must be a residual sequence of $\sum_{i=1}^{m} G_i$. Any central sequence in U(G) is equivalent to a central sequence whose elements lie in U(H), if H is canonically considered as a subalgebra of U(G), and H is strongly residual in G.

DEFINITION 2 ([4]). A sequence (T_n) in the unit sphere $(A)_1$ of $A = U(G)$ is an *ε-approximate* central sequence, if $\lim\sup \| [T_n, X] \|_2 < \epsilon$ for all $X \in (A)_1$. The set of all ε-approximate sequences is denoted by $C_A(\epsilon)$.

If H is strongly residual in G, then for all $(T_n) \in C_{U(G)}(\epsilon)$, there exists a sequence (T_n') in the unit sphere of U(H) such that $\lim\sup \| T_n - T_n' \|_2 < 14\epsilon$ (cf. [3], [5], [6]).

Let $G = (G_1, G_2, \ldots; H_1, H_2, \ldots)$ with $G_i = Z$ and let $Q(G, n) = \sum_{j=n}^{\infty} \oplus H_j$ and $Q(G, m, n) = \sum_{j=m}^{n} \oplus H_j$. Then $\{Q(G, n)\}$ is a residual sequence in G. Let $\{\Gamma_k : k = 1, 2, \ldots, m\}$ be a finite family with the form $\Gamma_k = (G_1, G_2, \ldots; H_1, H_2, \ldots)$ with $G_i = Z$.

Let $Q(\sum_{k=1}^{r} \oplus \Gamma_k, n) = \sum_{k=1}^{r} \oplus Q(\Gamma_k, n)$ be a residual sequence in G. This residual sequence is called the *canonical residual sequence.*

Let $Q(\sum_{k=1}^{r} \oplus \Gamma_k, n, m)$ denote $\sum_{k=1}^{r} \oplus Q(\Gamma_k, n, m)$.

A group G is called *of type 0* if it is commutative; G is called *of type i* if $G = \sum_{j=1}^{n} \oplus G_j = L_i$; G is called *of type i_∞* if $G = \sum_{j=1}^{\infty} \oplus G_j$ with $G_j = L_i$; G is called *of type $(i_1, i_2, \ldots, i_n)$* if $G = \sum_{j=1}^{n} \oplus G_j$, where G_j is of type i_j; G is called *of type $(i_1, i_2, \ldots, i_n)_\infty$* if $G = \sum_{j=1}^{n} \oplus G_j$ and some G_j are of type $i_{j\infty}$ and others are of type i_j.

Now let $U(G[I_1]) = A$ and $U(G[I_2]) = B$. Suppose that A is *-isomorphic to B. Then, under the identification $A = B$, we have two expressions $U(G[I_1])$ and $U(G[I_2])$.

We now assume that $A = B$ and reach a contradiction.

LEMMA 2 ([4]). For $\delta > 0$ and a positive integer n_1 there exists a positive integer n_2 such that $U(Q(G[I_2], n_2)) \overset{\delta}{\subset} U(Q(G[I_1], n_1))$.

Moreover, since $U(Q(G[I_i], n, n+1))$ is a factor, we have

LEMMA 3([4]). For a positive integer m_2 with $m_2 > n_2$, there exists a positive integer m_1 such that $m_1 > n_1$ and $U(Q(G[I_2], n_2, m_2)) \overset{9\delta}{\subset} U(Q(G[I_1], n_1, m_1))$.

Now let $I_1 = (p_i)$ and $I_2 = (q_j)$. Without loss of generality, we can assume that $q = q_1$. For $t = 10^{q_1}$, by applying Lemma 2 for I_1 and the symmetric form of Lemma 2 for I_2, we can choose positive integers $n_1, n_2, \ldots, n_t$ such that $n_2 < n_4 < n_6 < \ldots < n_t$, and

$$U(Q(G[I_2], n_t)) \overset{\delta}{\subset} U(Q(G[I_1], n_{t-1})) \overset{\delta}{\subset} \ldots \overset{\delta}{\subset} U(Q(G[I_2], n_2)) \overset{\delta}{\subset} U(Q(G[I_1], n_1)).$$

Then, by Lemma 3, we can choose positive integers $m_1, m_2, \ldots, m_t$ such that $m_2 > m_4 > \ldots > m_t$ and $m_1 > m_3 > \ldots > m_{t-1}$ with $m_t > n_t$ and

$$U(Q(G[I_2], n_t, m_t)) \overset{9\delta}{\subset} U(Q(G[I_1], n_{t-1}, m_{t-1})) \overset{9\delta}{\subset} \ldots \overset{9\delta}{\subset} U(Q(G[I_1], n_1, m_1)).$$

Since $Q(G[I_i], h, k)$ is a finite sum of the form $(G_1, G_2, \ldots; H_1, H_2, \ldots)$ with $G_i = Z$, it has the canonical residual sequence $\{Q(Q(G[I_i], h, k), n\}$. For simplicity we shall denote $Q(G[I_i], h, k)$ (resp. $Q(Q(G[I_i], h, k), n))$ by $Q_i(h, k)$ (resp. $Q_i^2((h, k), n)$).

LEMMA 4. Suppose $U(Q_1(h, k)) \overset{9\delta}{\subset} U(Q_2(i, j)) \overset{9\delta}{\subset} U(Q_1(p, q))$ with $h > p$ and $q > k$. Then for arbitrary positive integers r and w, there exists a positive integer s such that $U(Q_1^2(h, k), s)) \overset{(10)^3\delta}{\subset} U(Q_2^2(i, j), r)$ and $s > w$.

Proof. Suppose this is not true; then there exists $T_n \in (U(Q_1^2((h, k), n)))_1$ for each n with $n > w$ such that $\| T_n - S \|_2 \geqslant (10)^3\delta$ for all $S \in (U(Q_2^2((i, j), r))_1$.

Since $\{Q_1^2((h, k), n)\}$ is a residual sequence in $U(Q_1(h, k))$, (T_n) is a central sequence in $U(Q_1(h, k))$.

On the other hand, $Q_1(p, q) = Q_1(h, k) \oplus C$, where C is a subgroup of $Q_1(p, q)$; hence (T_n) is a central sequence in $U(Q_1(p, q))$. Now, take $T_n' \in (U(Q_2(i, j))_1$ such that $\| T_n - T_n' \|_2 < 9\delta$, and for arbitrary $X' \in (U(Q_2(i, j))_1$, take $X \in (U(Q_1(p, q))_1$ such that $\| X - X' \| < 9\delta$. Then

$$\| [X', T_n'] \|_2 = \| [X', T_n' - T_n] \|_2 + \| [T_n, X] \|_2 + \| [T_n, X - X'] \|_2$$

$$\leqslant 2\| T_n' - T_n \|_2 + \| [T_n, X] \|_2 + 2\| X - X' \|_2.$$

Hence $\limsup \| [X', T_n'] \|_2 \leqslant 18\delta + 18\delta < 37\delta$. Therefore there exists a sequence (T_n'') in $(U(Q_2^2((i, j), r)))_1$ such that $\limsup \| T_n' - T_n'' \|_2 < 14 \cdot 37\delta$. Then,

$$\| T_n - T_n'' \|_2 \leqslant \| T_n - T_n' \|_2 + \| T_n' - T_n'' \|_2 < 10^3\delta.$$

This is a contradiction and completes the proof.

Applying this lemma for I_1 and the symmetric one for I_2, we obtain positive integers $r_2, r_3, \ldots, r_t$ such that $r_2 < r_4 < \ldots < r_t$ and $r_3 < r_5 < \ldots < r_{t-1}$, and

$$U(Q_2{}^2((n_t, m_t), r_t)) \overset{10^3\delta}{\subset} \ldots \overset{10^3\delta}{\subset} U(Q_1{}^2((n_3, m_3), r_3)) \overset{10^3\delta}{\subset} U(Q_2{}^2((n_2, m_2), r_2)).$$

Now $Q(G[I_1], n, m) = \sum_{j=n}^{m} \oplus M_j(I_1)$ is of type $(p_1, p_2, \ldots, p_m)$. Then $Q_1{}^2((n,m), r)$ is of type $(p_1 - 1, \ldots, p_m - 1)$. Therefore, at this time, $Q_i{}^2((h, k), r)$ might contain a type 0 group as a direct summand.

We use the following notation: for $r < s$, $RQ_i{}^2((h, k), (r, s))$ = the center of $Q_i{}^2((h, k), r) \oplus (Q_i{}^2((h, k), r) \ominus Q_i{}^2((h, k), s + 1))$.

LEMMA 5. For arbitrary positive integers $s_t > r_t$, there exist positive integers $s_4, s_5, \ldots, s_t$ such that $s_4 > s_6 > \ldots > s_t$ and $s_5 > s_7 > \ldots > s_{t-1}$, and

$$U(RQ_2{}^2(n_t, m_t), (r_t, s_t))) \overset{10^5\delta}{\subset} \ldots \overset{10^5\delta}{\subset} U(RQ_1{}^2((n_5, m_5), (r_5, s_5)) \overset{10^5\delta}{\subset}$$
$$\overset{10^5\delta}{\subset} U(RQ_2{}^2((n_4, m_4), (r_4, s_4))).$$

Proof. $Q_2((n_{t-2}, m_{t-2})) = Q_2((n_t, m_t) \oplus H$, where H is a subgroup of $Q_2((n_{t-2}, m_{t-2}))$. Now consider $Q_2{}^2((n_t, m_t), r) \oplus H$ in $Q_2((n_{t-2}, m_{t-2}))$; then $Q_2{}^2((n_t, m_t), r)$ is strongly residual in $Q_2((n_t, m_t))$, and so $Q_2{}^2((n_t, m_t), r) \oplus H$ is strongly residual in $Q_2((n_{t-2}, m_{t-2}))$ for each r.

On the other hand

$$U(Q_1((n_{t-1}, m_{t-1}) \overset{9\delta}{\subset} U(Q((n_{t-2}, m_{t-2})) \overset{9\delta}{\subset} U(Q((n_{t-3}, m_{t-3}))).$$

Therefore by a method similar to the proof of Lemma 4, for each r there exists a k with $k - 1 > r_{t-1}$ such that $U(Q_1{}^2(n_{t-1}, m_{t-1})k) \overset{(10)^3\delta}{\subset} U(Q_2{}^2((n_t, m_t), r) \oplus \oplus H)$. Take $r = s_t + 1$; then for $T \in (U(RQ_2{}^2((n_t, m_t), (r_t, s_t))))_1 \subset (U(Q_2{}^2((n_t, m_t), r_t)))_1$, there exists $T' \in (U(Q_1{}^2((n_{t-1}, m_{t-1}), r_{t-1})))_1$ such that $\| T - T' \|_2 < < (10)^3\delta$. For $X' \in (U(Q_1{}^2((n_{t-1}, m_{t-1}), k)))_1$, take $X \in (U(Q_2{}^2((n_t, m_t), r) \oplus \oplus H))_1$ such that $\| X - X' \|_2 < 10^3\delta$; then

$$\| [T', X'] \|_2 \leqslant \| [T' - T, X'] \|_2 + \| [X' - X, T] \|_2 + \| [T, X] \|_2$$
$$< 2 \cdot 10^3\delta + 2 \cdot 10^3\delta,$$

because $[T, X] = 0$.

Thus there is a $T'' \in (U(Q_1{}^2((n_{t-1}, m_{t-1}), k))')_1 \cap U(Q_1{}^2((n_{t-1}, m_{t-1}), r_{t-1})$, where $(\ )'$ is the commutant of the W*-algebra $(\ )$, such that $\| T' - T'' \|_2 < 4 \cdot \cdot 4(10)^3\delta$ (cf. Lemma 4 in [4]). Hence $\| T - T'' \|_2 \leqslant \| T - T' \|_2 + \| T' - T'' \|_2 < < (10)^5\delta$. Clearly

$$U(Q_1{}^2((n_{t-1}, m_{t-1}), k)' \cap U(Q_1{}^2((n_{t-1}, m_{t-1}), r_{t-1})) =$$

$$= RQ_1{}^2((n_{t-1}, m_{t-1}), (r_{t-1}, k-1)).$$

Take $k-1$ as s_{t-1}. The remaining part is quite similar. This completes the proof.

Remark. The proof of Lemma 5 is due to B. Vowden.

$RQ_1{}^2((h,k),(i,j))$ is of type $(p_1-1, p_2-1, \ldots p_k-1)$ and $RQ_2{}^2((h,k), (i,j))$ is of type $(q_1-1, q_2-1, \ldots, q_k-1)$. They might contain a type 0 group as a direct summand. $RQ_1{}^2((h,k),(i,j)) = D \oplus W$, where D is the center of $RQ_i{}^2(h, k),(i,j))$, and W is of type $(i_1, i_2, \ldots, i_n)$ with $i_u \geqslant 1$ for $u = 1, 2, \ldots, n$.

Define the canonical residual sequence of $RQ_1{}^2((h,k),(i,j))$ as follows:

$$Q_1 RQ_1{}^2((h,k),(i,j),n) = D \oplus Q(W,n).$$

Quite similarly, we define the canonical residual sequence of $RQ_2{}^2$.

Now we shall continue this process by q_1 times. Then we have the following situation:

$$U(\Omega_t) \overset{k\delta}{\subset} U(\Omega_{t-1}) \overset{k\delta}{\subset} U(\Omega_{t-2}) \overset{k\delta}{\subset} U(\Omega_{t-3}),$$

where Ω_t contains a type 1 group as a direct summand and Ω_{t-1} does not; moreover $\Omega_{t-2} = \Omega_t \oplus R$, where R is a subgroup of Ω_{t-2}; K is a constant which does not depend on δ, and by the (q_1+1)-th process, we have

$$U(\Delta_t) \overset{K_1\delta}{\subset} U(\Delta_{t-1}) \overset{K_1\delta}{\subset} U(\Delta_{t-2}),$$

where K_1 does not depend on δ. Moreover, let $\Omega_t = E \oplus H$, where E is the center of Ω_t; then $\Delta_t = E \oplus E_1 \oplus W$, where E_1 is contained in the center of Δ_t, and $E_1 = Q(L_1, n)$ for some n.

On the other hand, the center of Δ_{t-1} is the same as the center C of Ω_{t-1}, because Ω_{t-1} does not contain a type 1 group as a direct summand.

LEMMA 6. For $X \in (U(E_1))_1$, there exists an element $X' \in (U(C))_1$ such that $\| X - X' \|_2 < 10^2 K_1 \delta$.

Proof. Put $X_n = X$; then (X_n) is a central sequence in (UE_1); it is a central sequence in $U(\Delta_{t-2})$ because $\Delta_{t-2} = \Delta_t \oplus \Gamma$ for some subgroup Γ. Let $Y' \in$ $\in (U(\Delta_{t-1}))_1$ such that $\| X - Y' \|_2 < K_1\delta$. Then by the arguments in the proof of Lemma 4, $\| [Y', W'] \|_2 < 5K_1\delta$ for all $W' \in (U(\Delta_{t-1}))_1$; hence there exists a central element X' of $(U(\Delta_{t-1}))_1$ such that $\| X' - Y' \|_2 \leqslant 2 \cdot 5K_1\delta$. Hence $\| X -$ $- X' \|_2 \leqslant \| X - Y' \|_2 + \| X' - Y' \|_2 < 10^2 K_1 \delta$. This completes the proof.

Now we shall prove Theorem 1.

Proof of Theorem 1. We have

$$(U(E_1))_1 \subset (U(L_1))_1 \subset (U(\Omega_t))_1 \overset{K\delta}{\subset} (U(\Omega_{t-1}))_1 .$$

By Lemma 6, for $X \in (U(E_1))_1$, there exists an $X' \in (U(C))_1$ such that $\| X - X' \|_2 < 10^2 K_1 \delta$. For arbitrary $Y \in (U(L_1))_1$, take $Y' \in (U(\Omega_{t-1}))_1$ such that $\| Y' - Y \|_2 < K\delta$. Then

$$\| [Y, X] \|_2 \leqslant \| [Y - Y', X] \|_2 + \| [Y', X - X'] \|_2 + \| [Y', X'] \|_2$$

$$\leqslant 2K\delta + 2 \cdot 10^2 K_1 \delta .$$

Hence there exists an element $X'' \in U(L_1) \cap U(L_1)' = (\lambda 1)$, where λ is a complex number, such that $\| X - X'' \|_2 < 4(K + 10^2 K_1)\delta$. We can choose δ arbitrarily small; hence $U(E_1)$ must be the center of $U(L_1)$.

On the other hand, $U(E_1)$ is not the center of $U(L_1)$ because $U(E_1) = U(\sum_{j=n}^{\infty} \oplus \bigcirc H_j)$ with $H_j = Z$. This is a contradiction and completes the proof.

Next we shall show the existence of an uncountable number of II_∞ factors.

Let F_2 be the free group on two generators g_1, g_2. Let S be the set of $g \in F_2$ which, when written as a power of g_1, g_2 of minimum length, end with a g_1^n, $n = \pm 1, \pm 2, \ldots$; then it is clear that $S \cup g_1 S g_1^{-1} = F_2 \backslash \{e\}$, $g_1 e g_1^{-1} = e$, and $g_2^n S g_2^{-n}$, $n = 0, \pm 1, \ldots$, are disjoint subsets of $F_2 \backslash \{e\}$, where e is the unit of F_2; therefore $\{e\}$ is strongly residual. Now let $R_j = F_2$ for $j = 1, 2, \ldots$ and $\Gamma = \sum_{j=1}^{\infty} \oplus R_j$. Put $\Gamma_n = \sum_{i=n}^{\infty} \oplus R_j$; then Γ_n is strongly residual in Γ, because Γ_n is strongly residual in Γ_n, and $\sum_{j=1}^{n-1} \oplus \{e_j\}$ is strongly residual in $\sum_{j=1}^{n-1} \oplus R_j$, where e_j is the unit of R_j; hence $\sum_{j=n}^{\infty} \oplus R_j$ is strongly residual in Γ.

Moreover $\Gamma_n \ominus \Gamma_{n+1} = R_n$ and $\bigcup_{n=1}^{\infty} \Gamma_n' = \sum_{j=1}^{\infty} \oplus R_j = \Gamma$; hence Γ_n is a residual sequence in Γ. Put $\Phi_n(I_1) = \Gamma_n \oplus Q(G[I_1], n)$ for $n = 1, 2, \ldots$; then $\{\Phi_n(I_1)\}$ is a residual sequence in $\Gamma \oplus G[I_1]$.

Next we shall prove

THEOREM 2. Let I_1 and I_2 be two subsets of I satisfying the conditions of Theorem 1; then $U(\Gamma \oplus G[I_1])$ is not *-isomorphic to $U(\Gamma \oplus G[I_2])$.

Proof. Since $\Phi_n(I_i) \ominus \Phi_{n+1}(I_i) = (\Gamma_n \ominus \Gamma_{n+1}) \oplus (Q(G[I_i], n) \ominus Q(G[I_i], n + L))$. $U(\Phi_n(I_i) \ominus \Phi_{n+1}(I_i))$ is a factor for $n = 1, 2, \ldots$ and $i = 1, 2$.

Therefore we can apply the lemmas of McDuff [4]. Now suppose that $U(\Gamma \oplus \oplus G[I_1]) = U(\Gamma \oplus G[I_2])$; then we have situations which are similar to Lemmas 2 and 3 for two residual sequences $\{\Phi_n(I_i)\}$, $i = 1, 2$.

On the other hand, $\Gamma_{mn} = \Gamma_m \ominus \Gamma_n$, $m < n$, has the strong residual subgroup $\{e\}$; hence we have the same relations as in the previous case:

$$U(Q_2{}^2((n_t, m_t), r_t) \overset{10^3\delta}{\subset} \dots \overset{10^3\delta}{\subset} U(Q_1{}^2((n_3, m_3), r_3) \overset{10^3\delta}{\subset} U(Q_2{}^2((n_2, m_2), r_2)).$$

This is a contradiction and completes the proof.

THEOREM 3. Suppose that I_1, I_2 satisfy the conditions of Theorem 1, and let B be a type I factor; then $B \otimes U(\Gamma \oplus G[I_1])$ is not *-isomorphic to $B \otimes U(\Gamma \oplus G[I_2])$.

Proof. $B \otimes U(\Gamma \oplus G[I_i]) = B \otimes U(\Gamma) \otimes U(G[I_i]) = B \otimes U(G[I_i]) \otimes \bigotimes U(R_n)$, where $\bigotimes (UR_n)$ is the canonical infinite tensor product of $\{U(R_n)\}$ (cf. [6]); hence $B \otimes U(\Gamma \oplus G[I_i]) \otimes A$ is *-isomorphic to $B \otimes U(\Gamma \oplus G[I_i])$, $i = 1, 2$, where A is the hyperfinite II_1 factor (cf. [6]). We shall denote $U(\Gamma \oplus G[I_1])$ by N. Let φ be a normal, faithful semi-finite trace on B, and let τ_1 (resp. τ_2) be the normalized trace on N (resp. A); then $\varphi \otimes \tau_1 \otimes \tau_2$ will define a normal, faithful semi-finite trace on $B \otimes N \otimes A$. Now, let E be a minimal projection of B; then $E \otimes 1_N$ is a finite projection in $B \otimes N$, where 1_N is the unit of N; moreover $(E \otimes 1_N)B \otimes N(E \otimes 1_N) = E \otimes N$; hence it is *-isomorphic to N. For arbitrary positive a with $a \leqslant \varphi \otimes \tau_1 \otimes \tau_2 (E \otimes 1_N \otimes 1_A)$, we have a projection P in A such that $\varphi \otimes \tau 1 \otimes \tau_2 (E \otimes 1_N \otimes P) = a$, where 1_A is the unit of A. Now suppose that $B \otimes N \otimes A$ is *-isomorphic to $B \otimes U(\Gamma \oplus G[I_2]) \otimes A$; then there exists a finite projection E_1 in $B \otimes N \otimes A$ such that $E_1(B \otimes N \otimes A)E_1$ is *-isomorphic to $U(\Gamma \oplus G[I_2]) \otimes A$.

Take $P_0 \in A$ such that $n_0\varphi \otimes \tau_1 \otimes \tau_2(E \otimes 1_N \otimes P_0) = \varphi \otimes \tau_1 \otimes \tau_2 (E_1)$ for some positive integer n_0. Then there exists a family $\{E_{1,i}: i = 1, 2, \dots, n_0\}$ of mutually orthogonal, equivalent projections in $B \otimes N \otimes A$ such that

$$E_{1,i} \sim E \otimes 1_N \otimes P_0, \; E_{1,i} \leqslant E_1 \text{ and } \sum_{i=1}^{n_0} E_{1,i} = E_1.$$

Since $E_{1,i} \sim E \otimes 1_N \otimes P_0$, $E_{1,i}(B \otimes N \otimes A)E_{1,i}$ is *-isomorphic to $(E \otimes 1_N \otimes P_0)(B \otimes N \otimes A)(E \otimes 1_N \otimes P_0)$. On the other hand, $(E \otimes 1_N \otimes P_0)(B \otimes N \otimes A)(E \otimes 1_N \otimes P_0) = E \otimes N \otimes P_0AP_0$, and since P_0AP_0 is *-isomorphic to A, $E \otimes N \otimes P_0AP_0$ is *-isomorphic to $N \otimes A$. Since $E_1(B \otimes N \otimes A)E_1$ is *-isomorphic to $E_{1,i}(B \otimes N \otimes A)E_{1,i} \otimes B_{n_0}$, it is *-isomorphic to $N \otimes A$, and B_{n_0} is a type I_{n_0} factor. Hence $U(\Gamma \oplus G[I_2]) \otimes A$ is *-isomorphic to $N \otimes A$.

Since $U(\Gamma \oplus G[I_i]) \otimes A$ is *-isomorphic to $U(\Gamma \oplus G[I_i])$, $i = 1, 2$, we have a contradiction. This completes the proof.

As a corollary, we have

COROLLARY 2. There exists an uncountable number of II_∞ factors on a separable Hilbert space.

REFERENCES

1. W. Ching, Non-isomorphic non-hyperfinite factors, to appear, Canadian J. Math.
2. J. Dixmier and E. C. Lance, Deux nouve facteurs de type II_1, to appear, Inventiones Math.
3. F. Murray and J. von Neumann, On rings of operators, IV, Ann. of Math., 44 (1943), pp. 716-808.
4. D. McDuff, A countable infinity of II_1 factors, to appear.
5. L. Pukansky, Some examples of factors, Publ. Math. Debrecen, 4 (1956), pp. 135-156.
6. S. Sakai, Asymptotically abelian II_1 factors, to appear.
7. J. Schwartz, Two finite, non-hyperfinite, non-isomorphic factors, Comm. Pure Appl. Math., 16 (1963), pp. 19-26.
8. G. Zeller-Meier, Deux autres facteurs de type II_1, to appear, Inventiones Math.
9. D. McDuff, The revised form of [4].
10. D. McDuff, Uncountably many II_1 factors, to appear.
11. R. Powers, Representations of uniformly hyperfinite algebras and their associated von Neumann algebras, Ann. of Math., 86 (1967), pp. 138-171.

AN UNCOUNTABLE FAMILY OF NON-HYPERFINITE TYPE III FACTORS

Shôichirô Sakai[1] – *University of Pennsylvania*

§1. INTRODUCTION. In the present paper, by using the tensor products of the uncountable family of type II_1 factors in [16] and a type III factor in [5], we shall construct an uncountable family of non-hyperfinite type III factors.

Throughout this paper we shall use the same notations as in [16].

§2. CONSTRUCTION OF EXAMPLES. First of all we shall extend the notion of a central sequence to arbitrary W*-algebras.

DEFINITION 1. Let M be a W*-algebra, and let (x_n) be a uniformly bounded sequence in M. Then (x_n) is called a *central sequence* if $[x_n, x] \to 0$ (strongly) for all $x \in M$, where $[x_n, x] = x_n x - x x_n$.

DEFINITION 2. Let (x_n), (y_n) be two central sequences in M. Then (x_n) is said to be *equivalent* to (y_n) if $x_n - y_n \to 0$ (strongly); we shall denote the equivalence of (x_n) and (y_n) by $(x_n) \sim (y_n)$.

Pukansky [5] proved that there exists a type III factor P in a separable Hilbert space $\mathcal{H}$, as follows: P has a separating, generating vector ξ_0, ($\|\xi_0\| = 1$), and there exist two elements a_1, a_2 in P and a fixed positive number K such that for $a \in P$, $\| [a, a_i] \|_2^2 \leqslant \epsilon$ $(i = 1, 2)$ imply $|(a\xi_0, \xi_0)|^2 \geqslant \| a \|_2^2 - K\epsilon$, where $\| x \|_2 = \| x\xi_0 \|$, and ϵ is an arbitrary positive number.

Let G be a discrete group with unit e, and let U(G) be the W*-algebra generated by the left regular representation of G. Suppose that U(G) is a factor. Then the tensor product $P \otimes U(G)$ is a type III factor [15]. For $g \in G$, let $\mathcal{H}_g$ be an $\aleph_0$-dimensional Hilbert space and J_g an isometric linear mapping of $\mathcal{H}$ onto $\mathcal{H}_g$. Let $\tilde{\mathcal{H}}$ be the direct sum of $\mathcal{H}_g$ for $g \in G$. Then $P \otimes U(G)$ is *-isomorphic to a W*-algebra A_G on $\tilde{\mathcal{H}}$ such that every element $S \in A_G$ can be represented by a matrix of the form $(T_{st^{-1}})$ with $T_g \in P$ for all $g \in G$ (cf. [12], p. 131).

We shall identify $P \otimes U(G)$ with A_G. Define a normal state φ_G on A_G as follows: $\varphi_G((T_{st^{-1}})) = (T_e\xi_0, \xi_0)$. Let $A_{i,e} = a_i$ and $A_{i,g} = 0$ for $g \neq e$ $(i = 1, 2)$; then $[(A_{i,st^{-1}}), (T_{st^{-1}})] = ([a_i, T_{st^{-1}}])$.

[1] This research supported by NSF.

LEMMA 1. For every central sequence in $P \otimes U(G)$, there exists an equivalent central sequence in $1 \otimes U(G)$.

Proof. Put $|||(T_{st}-1)|||_2 = \varphi_G((T_{st}-1)^*(T_{st}-1)) = (\sum_{g\in G} \| T_g\xi_0 \|^2)^{1/2}$. Let $x = (T_{st}-1)$ and suppose that $|||[(A_{i,st}-1),(T_{st}-1)]|||_2^2 = |||([a_i, T_{st}-1])|||_2^2 = \sum_{g\in G} \| [a_i, T_g] \|_2^2 < \epsilon$. Then $|(T_g\xi_0, \xi_0)|^2 \geq \| T_g \|_2^2 - K(\| [a_1, T_g] \|_2^2 + \| [a_2, T_g] \|_2^2)$. Therefore, $|||(T_{st}-1) - ((T_{st}-1\xi_0, \xi_0)_1)|||_2^2 = \sum_{g\in G} \| T_g - (T_g\xi_0, \xi_0)_1 \|_2^2 = \sum_{g\in G} (\| T_g\xi_0 \|_2^2 - |(T_g\xi_0, \xi_0)|^2 \leq \sum_{g\in G} K(\| [a_1, T_g] \|_2^2 + \| [a_2, T_g] \|_2^2) < 2K\epsilon$.

Now we shall show that $((T_{st}-1\xi_0, \xi_0)_1)$ is a bounded operator which belongs to $1 \otimes U(G)$. Put $\psi_0(x) = (x\xi_0, \xi_0)$ for $x \in P$ and $f \in U(G)_*$, where $U(G)_*$ is the predual of $U(G)$.

Then $\psi_0 \otimes f$ is a normal linear functional of $P \otimes U(G)$; moreover $|\psi_0 \otimes f((T_{st}-1)| \leq \|f\| \, \|(T_{st}-1)\|$ for $f \in U(G)_*$; hence there exists an element d in $U(G)$ such that $\psi_0 \otimes f((T_{st}-1)) = f(d)$ and $\| d \| \leq \| (T_{st}-1) \|$. Let $f(y) = (y\epsilon_g, \epsilon_k)$ for $y \in U(G)$, where $\epsilon_g(k) = \delta_{g,k}$ and $\epsilon_h(k) = \delta_{h,k}$ for $k \in G$, where δ is the Kronecker symbol. Then $\psi_0 \otimes f((T_{st}-1)) = ((T_{st}-1)\xi_0 \otimes \epsilon_g, \xi_0 \otimes \epsilon_h) = (T_{hg^{-1}}\xi_0, \xi_0)$.

On the other hand $f(d) = (d\epsilon_g, \epsilon_h) = d(hg^{-1})$; hence $(T_g\xi_0, \xi_0) = d(g)$ for $g \in G$, and so $((T_{st}-1\xi_0, \xi_0)_1)$ belongs to $1 \otimes U(G)$, and $\|(T_{st}-1\xi_0, \xi_0)_1\| \leq \| (T_{st}-1) \|$. This completes the proof.

Now let $I_1 = (p_1, p_2, \ldots)$ and $I_2 = (q_1, q_2, \ldots)$ be two sequences of positive integers, and let $G[I_1]$ and $G[I_2]$ be the discrete groups constructed in [16]. Then we shall prove the following theorem.

THEOREM. Suppose that I_2 contains a positive integer q such that $q \notin I_1$. Then $P \otimes U(G[I_1])$ is not *-isomorphic to $P \otimes U(G[I_2])$. In particular, they are not *-isomorphic if $I_1 = I_2$ as a set of positive integers.

In order to prove the theorem, we first provide some preliminary considerations.

Let $A = P \otimes U(G[I_1])$ and $B = P \otimes U(G[I_2])$. Suppose that A is *-isomorphic to B. Then under the identification $A = B$, we have two expressions $P \otimes U(G[I_1])$ and $P \otimes U(G[I_2])$. Pick the expression A and take the normal state $\varphi_{G[I_1]}$ and define $|||x|||_2 = \varphi_{G[I_1]}(x^*x)$ for $x \in A$. Then the topology of this metric is equivalent to the strong operator topology on bounded spheres.

We shall extend the notion $F \overset{\delta}{\subset} E$ in A, using the norm $|||\cdot|||_2$ instead of the trace norm in finite algebras. We can easily see that $\varphi_{G[I_1]}(xy) = \varphi_{G[I_1]}(yx)$ for $x \in P \otimes U(G[I_1])$ and $y \in 1 \otimes U(G[I_2])$. Now let (u_n) be a central sequence in $U(G[I_1])$ such that $u_n^*u_n = 1$; then we shall show that $(1 \otimes u_n)$ is a central sequence in $P \otimes U(G[I_1])$. For $x \in P \otimes U(G[I_1])$ and $\epsilon > 0$, there exists an element x_0 in the algebraic tensor product $P \odot U(G[I_1])$ such that $|||x - x_0|||_2 < \epsilon$. Then $|||(1 \otimes u_n)^*x(1 \otimes u_n) - x|||_2 \leq |||1 \otimes u_n^*x1 \otimes u_n - 1 \otimes u_n^*x_0 1 \otimes u_n|||_2 + |||1 \otimes u_n^*x_0 1 \otimes u_n - x_0|||_2 + |||x - x_0|||_2$.

Hence $\limsup |||(1 \otimes u_n)^*x(1 \otimes u_n) - x|||_2 \leqslant 2\,|||x - x_0|||_2 < 2\epsilon$.

Hence $[x, 1 \otimes u_n] \to 0$ (strongly).

Now conisder the sequence $\{1 \otimes U(Q(G[I_1], n))\}$ of W*-subalgebras in $P \otimes \otimes U(G[I_1])$. Take $y_n \in 1 \otimes U(Q(G[I_1], n))$ with $\| y_n \| \leqslant 1$, $y_n^* = y$; then

$$y_n = [y_n + i\sqrt{1 - y_n^2} + (y_n - i\sqrt{1 - y_n^2})]/2;$$

$y_n \pm i\sqrt{1 - y_n^2}$ are unitary; since the sequences $(y_n \pm i\sqrt{1 - y_n^2}$ are central in $1 \otimes \otimes U(G[I_1])$, they are central in $P \otimes U(G[I_1])$, hence (y_n) is central in $P \otimes U(G[I_1])$. Now we can easily see that every uniformly bounded sequence (x_n) with $x_n \in 1 \otimes \otimes U(Q(G[I_1], n))$ is central. Conversely, by Lemma 1 every central sequence is equivalent to a central sequence in $1 \otimes U(G[I_i])$ $(i = 1, 2)$.

For simplicity, we shall identify $1 \otimes U(G[I_i])$ with $U(G[I_i])$.

LEMMA 2. For every positive number δ with $0 < \delta < 1$ and every positive integer n_1, there exists a positive integer n_2 such that

$$U(Q(G[I_2], n_2)) \overset{\delta}{\subset} U(Q(G[I_1], n_1)).$$

Proof. Suppose that this is not true; then there exists $T_n \in (U(Q(G[I_2], n)))_1$ for each n such that $|||T_n - S|||_2 \geqslant \delta$ for all $S \in (U(Q(G[I_1], n_1)))_1$. Since $\{Q(G[I_2], n)\}$ is a residual sequence in $G[I_2]$, (T_n) is a central sequence in $P \otimes U(G[I_1])$; therefore there exists a sequence (T_n') in $(U(Q(G[I_1], n_1)))_1$ such that $|||T_n - - T_n'|||_2 \to 0$ $(n \to \infty)$. This is a contradiction and completes the proof.

LEMMA 3. For every positive integer m_2 with $m_2 > n_2$, there exists a positive integer m_1 such that $m_1 > n_1$ and

$$U(Q(G[I_2], n_2, m_2)) \overset{9\delta^{1/2}}{\subset} U(Q(G[I_1], n_1, m_1)).$$

Proof. For $T \in U(Q(G[I_2], n_2, m_2)))_1$ take $T' \in U(Q(G[I_1], n_1))_1$ such that $|||T - T'|||_2 < \delta$. Now by Lemma 2, we can choose a positive integer r such that $r - - 1 > n_1$ and $U(Q(G[I_1], r)) \overset{\delta}{\subset} U(Q(G[I_2], m_2 + 1))$. For $X' \in (U(Q(G[I_1], r)))_1$, take $X \in (U(Q(G[I_2], m_2 + 1)))_1$ such that $|||X' - X|||_2 < \delta$. Then

$$\begin{aligned} |||[X', T']|||_2 &\leqslant |||[X', T' - T]|||_2 + |||[X, T]|||_2 + |||[X - X', T]|||_2 \\ &\leqslant |||[X', T' - T]|||_2 + |||[X - X', T]|||_2, \end{aligned}$$

because $[X, T] = 0$. Since $\varphi_{G[I_1]}(xy) = \varphi_{G[I_1]}(yx)$ for $y \in U(G[I_1])$ and $x \in P \otimes \otimes U(G[I_1])$,

$$\begin{aligned}|||[X', T'-T]|||_2 &\leq |||X'(T'-T)|||_2 + |||(T'-T)X'|||_2 \\ &\leq |||T'-T|||_2 + \varphi_{G[I_1]}(X'^*(T'-T)^*(T'-T)X')^{1/2} \\ &< \delta + \varphi_{G[I_1]}(X'X'^*(T'-T)^*(T'-T))^{1/2} \\ &< \delta + \varphi_{G[I_1]}(X'X'^*(T'-T)^*(T'-T)X'X'^*)^{1/4} \cdot \\ &\quad \cdot \varphi_{G[I_1]}((T'-T)^*(T'-T))^{1/4} \\ &< \delta + 4^{1/4}\delta^{1/2} < 3\delta^{1/2}.\end{aligned}$$

Moreover,

$$\begin{aligned}|||[X'-X, T]|||_2 &\leq |||(X'-X)T\|_2 + |||T(X'-X)|||_2 \\ &< \delta + |||(X'-X)T|||_2 = \delta + \varphi_{G[I_1]}(T^*(X'-X)^*(X'-X)T)^{1/2}.\end{aligned}$$

Now we show that for a W*-subalgebra C of $P \otimes U(G[I_1])$ such that $C \overset{\delta}{\subset} U(G[I_1])$, we have

$$|\varphi_{G[I_1]}(xz) - \varphi_{G[I_1]}(zx)| < 4\delta$$

for $x \in (P \otimes U(G[I_1]))_1$ and $z \in (C)_1$. Suppose $z^* = z$, and take $y \in (U(G[I_1]))_1$ such that $|||z - y|||_2 < \delta$; then

$$\begin{aligned}|\varphi_{G[I_1]}(xz) - \varphi_{G[I_1]}(zx)| &\leq |\varphi_{G[I_1]}(x(z-y))| + |\varphi_{G[I_1]}((z-y)x)| \cdot \\ &\quad \cdot |\varphi_{G[I_1]}(xy - yx)| \\ &\leq \varphi_{G[I_1]}(xx^*)^{1/2}\varphi_{G[I_1]}((z-y)^*(z-y)^{1/2} + \\ &\quad + \varphi_{G[I_1]}(z-y)(z-y)^*)^{1/2}\varphi_{G[I_1]}(x^*x)^{1/2} \\ &\leq |||z-y|||_2 + |||z-y^*|||_2.\end{aligned}$$

On the other hand,

$$\begin{aligned}|||z-y^*|||_2^2 &= \varphi_{G[I_1]}((z-y)(z-y)^*) = \varphi_{G[I_1]}(z^2 - yz - zy^* + yy^*) \\ &= \varphi_{G[I_1]}(z^2 - zy - y^*z + y^*y) = |||z-y|||_2^2.\end{aligned}$$

Hence

$$|\varphi_{G[I_1]}(xz) - \varphi_{G[I_1]}(zx)| < 2\delta.$$

Therefore for arbitrary $z \in (C)_1$ we have

$$|\varphi_{G[I_1]}(xz) - \varphi_{G[I_1]}(zx)| < 4\delta.$$

Hence

$$|\varphi_{G[I_1]}(T^*(X' - X)^*(X' - X)T) - \varphi_{G[I_1]}(TT^*(X' - X)^*(X' - X)) < 4\delta.$$

On the other hand,

$$|\varphi_{G[I_1]}(TT^*(X' - X)^*(X' - X))| \leqslant 2|||X' - X|||_2.$$

Hence $\varphi_{G[I_1]}(T^*(X' - X)^*(X' - X)T)^{1/2} < 3\delta^{1/2}$; therefore

$$|||[X - X', T]|||_2 < 4\delta^{1/2} \text{ and } |||[X', T']|||_2 < 7\delta^{1/2}.$$

Now for any unitary element U' in $U(Q(G[I_1], r))$, we have $|||U'^*T'U' - T'|||_2 < 7\delta^{1/2}$. Let K be the strongly closed convex set generated by $\{U'^*T'U'\}$; then K contains a unique element $T'' \in U(Q(G[I_1], n_1, r))' \cap U(Q(G[I_1], n_1) = U(Q(G[I_1], n_1, r-1))$; hence $|||T'' - T'|||_2 < 8\delta^{1/2}$. Then $|||T - T''|||_2 \leqslant |||T - T'|||_2 + |||T' - T''|||_2 < \delta + 8\delta^{1/2} < < 9\delta^{1/2}$. This completes the proof.

LEMMA 4. Suppose that $U(Q(G[I_1], n_3)) \subset U(Q(G[I_2], n_2))$. For every positive integer m_3 with $m_3 > n_3$, there exists a positive integer m_2 such that $m_2 > n_2$ and $U(Q(G[I_1], n_3, m_3)) \overset{9\delta^{1/2}}{\subset} U(Q(G[I_2], n_2, m_2))$.

Proof. By Lemma 2, we can choose a positive integer s such that $s - 1 > n_2$ and $U(Q(G[I_2], s)) \overset{\delta}{\subset} U(Q(G[I_1], m_3 + 1))$. For $T' \in (U(Q(G[I_1], n_3, m_3)))_1$, take $T \in (U(Q(G[I_2], n_2)))_1$ such that $|||T' - T|||_2 < \delta$. For $X \in (U(Q(G[I_2], s)))_1$, take $X' \in (U(Q(G[I_1], m_3 + 1)))_1$ such that $|||X - X'|||_2 < \delta$. Then

$$|||[X, T]|||_2 < |||[X, T - T']|||_2 + |||[X', T']|||_2 + |||[X - X', T']|||_2.$$

Since $X \in U(Q(G[I_2], s)) \overset{\delta}{\subset} U(G[I_1])$ and $T' \in U(G[I_1])$, by arguments similar to the proof of Lemma 3, we have $|||[X, T]|||_2 < 7\delta^{1/2}$. Hence the other part of the proof is quite similar to the proof of Lemma 3.

The remaining part of the proof of the lemma is an analogous modification of the corresponding part of the proof of Theorem 2 in [16]. Then finally we have the following situation: $D \overset{R\delta^a}{\subset} (\lambda 1)$ and $(\lambda 1) \subset D$, where D is an infinite-dimensional commutative W*-algebra, $(\lambda 1)$ is one-dimensional, and R, a are positive numbers which do not depend on δ. This is a contradiction and completes the proof.

COROLLARY. There exists an uncountable family of non-isomorphic, non-hyperfinite type III factors in a separable Hilbert space.

Proof. It is enough to show that $P \otimes U(G[I_1])$ is not hyperfinite. Suppose that it is hyperfinite; then by [7] there exists a norm-one projection of $B(\mathcal{H} \otimes \ell_2(G[I_1]))$ onto $(P \otimes U(G[I_1]))' = P' \otimes U(G[I_1])'$ [17], where $B(\mathcal{H} \otimes \ell_2(G[I_1]))$ is the W*-algebra of all bounded operators on $\mathcal{H} \otimes \ell_2(G[I_1])$. On the other hand, there exists a norm-one projection of $P' \otimes U(G[I_1])'$ onto $1 \otimes U(G[I_1])'$ [15]; hence there exists a norm-one projection of $B(\mathcal{H} \otimes \ell_2(G[I_1]))$ onto $1 \otimes U(G[I_1])'$. But $U(G[I_1])'$ is *-isomorphic to $U(T[I_1])$, and $G[I_1]$ contains a free group on two generators as a subgroup, a contradiction ([7], [13], [14]). This completes the proof.

REFERENCES

1. W. Ching, Non-isomorphic, non-hyperfinite factors, to appear, Canadian J. Math.
2. J. Dixmier and E. C. Lance, Deux nouve facteurs de type II_1, to appear, Inventiones Math.
3. F. Murray and J. von Neumann, On rings of operators, IV, Ann. of Math., 44 (1943), pp. 716-808.
4. D. McDuff, A countable infinity of II_1 factors, to appear.
5. L. Pukansky, Some examples of factors, Publ. Math. Debrecen, 4 (1956), pp. 135-156.
6. S. Sakai, Asymptotically abelian II_1 factors, to appear.
7. J. Schwartz, Two finite, non-hyperfinite, non-isomorphic factors, Comm. Pure Appl. Math., 16 (1963), pp. 19-26.
8. G. Zeller-Meier, Deux autres facteurs de type II_1, to appear, Inventiones Math.
9. D. McDuff, The revised form of [4].
10. D. McDuff, Uncountably many II_1 factors, to appear.
11. R. Powers, Representations of uniformly hyperfinite algebras and their associated von Neumann algebras, Ann. of Math., 86 (1967), pp. 138-171.
12. J. Dixmier, Les algebras d'operateurs dans l'espace hilbertien, Paris, Gauthier-Villars, 2nd ed., 1969.
13. J. Hakeda and J. Tomiyama, On some extension properties of von Neumann algebras, Tohoku Math. J., (2), 19 (1967), pp. 315-323.
14. S. Sakai, On the hyperfinite II_1 factor, Proc. Amer. Math. Soc., 19 (1968), pp. 589-591.
15. S. Sakai, The theory of W*-algebras, Lecture Notes, Yale University, 1962.
16. S. Sakai, An uncountable number of II_1, II_∞ factors, to appear, J. of Functional Anal.
17. S. Sakai, On the tensor product of W*-algebras, Amer. J. Math., 90 (1968), pp. 335-341.

THE ISOMORPHISM PROBLEM FOR MEASURE-PRESERVING TRANSFORMATIONS

Donald Ornstein – *Stanford University*

In this talk we will concern ourselves with the problem of classifying the measure-preserving, invertible transformations of the unit interval. In fact we will concern ourselves with the classification of the simplest examples of such transformations, namely the Bernoulli shifts.

A Bernoulli shift can be described as follows: let S be a set with a finite or countable number of points, where the ith point is assigned measure p_i and $\Sigma p_i = 1$. Let X be the product of a doubly infinite sequence of copies of S, and put the product measure on X. Let $\{\ldots, x_{-1}, x_0, x_1, \ldots\}$ be a point in X. Define $T\{x_i\} = \{y_i\}$ where $y_{i+1} = x_i$ (that is, T shifts every sequence).

A Bernoulli shift is the simplest example of an ergodic (the only invariant sets have measure 0 or 1), invertible, measure-preserving transformations in the following sense: any ergodic, invertible, measure-preserving transformation can be represented (except, of course, for sets of measure 0) in the above form if, instead of putting the product measure on X, we put some other measure invariant under T.

We will say that T_1 acting on X_1 is isomorphic to T_2 acting on X_2 if there are subsets $X_1' \subset X_1$ and $X_2' \subset X_2$ of measure 1 and invariant under T_1 and T_2, respectively, and if there is an invertible, measure-preserving transformation T mapping X_1' onto X_2' such that if x is in X_1', then $TT_1(x) = T_2T(x)$.

There is another formulation of the problem which I believe brings out more clearly the nature of the problem. We will say that an invertible, measure-preserving transformation T on the unit interval (or a Lebesgue space) is a Bernoulli shift if there is a partition P of X consisting of a finite or countable number of sets P_i such that:

(1) the T^iP are independent (that is, $m(\bigcap_{-n}^{n} T^iP_{f(i)}) = \prod_{-n}^{n} m(P_{f(i)})$, where m denotes the measure of a set);

(2) the T^iP generate the full σ-algebra of X (that is, if E is a measurable set, then for each ϵ we can find an n and a set $\tilde{E}$ in the algebra generated by T^iP, $-n \leqslant i \leqslant n$, such that the measure of the symmetric difference between E and $\tilde{E}$ is less than ϵ).

T is isomorphic to a Bernoulli shift in our previous sense. (Furthermore, the sets P_i would correspond the set of all points whose 0th coordinate is the ith point in S.)

Let T be a transformation on X and P a partition such that the T^iP are independent and generate. Let $\bar{T}$ be a transformation on $\bar{X}$ and $\bar{P}$ a partition such that the $\bar{T}^i\bar{P}$ are independent and generate. The question of whether T and $\bar{T}$ are isomorphic comes to the following. Can we find a partition $\tilde{P}$ of $\bar{X}$ such that the ith set in $\tilde{P}$ has the same measure as the ith set in P and the $\bar{T}^i\tilde{P}$ are independent and generate the full σ-algebra of $\bar{X}$?

Halmos, in his Lectures on Ergodic Theory, pointed out that it was ~~now~~ not known if the 2-shift (P has 2 sets, each of measure $\frac{1}{2}$) is isomorphic to the 3-shift (P has 3 sets, each of measure $\frac{1}{3}$). One felt that there was an important gap in our understanding of measure-preserving transformations if we couldn't decide whether the two simplest examples were the same.

In 1958, Kolmogorov made one of the most important advances in ergodic theory by introducing a new invariant, called *entropy*. If T is a Bernoulli shift whose independent generator P has sets P_i of measure p_i, then the entropy of T is $-\Sigma p_i \log p_i$. Since $\frac{1}{2}\log\frac{1}{2} + \frac{1}{2}\log\frac{1}{2} \neq \frac{1}{3}\log\frac{1}{3} + \frac{1}{3}\log\frac{1}{3} + \frac{1}{3}\log\frac{1}{3}$, the 2-shift is not isomorphic to the 3-shift.

There are still many Bernoulli shifts with the same entropy and the question remained: which of these are isomorphic? Was it possible that two Bernoulli shifts were isomorphic only if they were identical (that is, the $\bar{p}_i$ were simply a rearrangement of the p_i)? Mesalkin ruled out this possibility by showing that if T and $\bar{T}$ had the same entropy, and if all of the p_i and $\bar{p}_i$ were powers of a simple rational number, then T and $\bar{T}$ were isomorphic.

Sinai made substantial progress toward classifying Bernoulli shifts by proving the following theorem. Let T be a Bernoulli shift on X with a partition P, the measure of whose i^{th} set is p_i, and the T^iP are independent and generate. Let $\bar{T}$ be a Bernoulli shift on $\bar{X}$ with a partition $\bar{P}$, the measure of whose i^{th} set is $\bar{p}_i$, and the $\bar{T}^i\bar{P}$ are independent and generate. Assume that $\Sigma p_i \log p_i = \Sigma \bar{p}_i \log \bar{p}_i$ (that is, T and $\bar{T}$ have the same entropy). Then we can find a partition $\tilde{P}$ of X, the measure of whose i^{th} set is $\bar{P}_i$, and the $T^i\tilde{P}$ are independent. The $T^i\tilde{P}$ do not necessarily generate. If they did, this would have shown that T and $\bar{T}$ are isomorphic.

We now know that any two Bernoulli shifts with the same entropy are isomorphic [1]. The method used here can be modified to give various other results, and I would now like to discuss some of these.

In [1], we only consider the case where P has a finite number of elements. Smorodinsky showed [5] that the argument in [1] could be modified to include the case where P is countably infinite and $\Sigma -p_i \log p_i < \infty$. In [2] we show that any two Bernoulli shifts for which $\Sigma -p_i \log p_i = \infty$ are isomorphic. Actually, in [2] we prove a little more, and to state this result we will first define a generalized Bernoulli shift as follows. Let S be a Lebesgue measure space of total measure 1. Let X be the product of a doubly infinite sequence of copies of S. For our measure on X we will take the product measure. We define T, as before, to be the shift operator. Our result is that any two generalized Bernoulli shifts with the same (finite or infinite) entropy are isomorphic. [Note that if the measure on S has a continuous part, then the entropy is infinite. Otherwise, S has a countable number of points (or, after throwing away a set of measure 0, S has a countable number of points).]

factor

The method used in [1] can also be modified to show that certain transformations are Bernoulli shifts.

In [3], we show that a subshift of a Bernoulli shift with finite entropy is a Bernoulli shift. By this we mean the following. Let T be a Bernoulli shift with finite entropy. Let A be a σ-algebra (of measurable sets) invariant under T. We can then find a finite partition P whose sets are in A such that the T^iP are independent and the T^iP generate A.

In [4] (a joint paper with N. Friedman), we modify the method in [1] to show that Markov shifts are Bernoulli shifts. By a Markov shift we mean the following. As before, we assume there is a countable partition P (with sets P_i of measure p_i) such that the T^iP generate. We do not, however, assume that the T^iP are independent. Instead we assume that

$$m\left(\bigcap_0^n T^iP_{f(i)}\right)/m\left(\bigcap_0^{n-1} T^iP_{f(i)}\right) = m\left(\bigcap_{n-1}^n T^iP_{f(i)}\right)/m(T^{n-1}P_{f(n-1)}).$$

This says that the distribution of T^nP given T^iP, $i = 0, \ldots, n-1$ depends only on $T^{n-1}P$. (Another way to say this is to say that a Markov shift is a shift in the space of realization of a Markov chain with finite invariant measure.)

If we go back k steps instead of one step, we get a generalized Markov shift, which is again a Bernoulli shift.

Various transformations that arise in other contexts can be shown to be Markov shifts. For example, Sinai has shown that a wide class of automorphisms of the n-dimensional torus are Markov shifts. Previously, Adler and Weiss showed that the mixing automorphisms of the 2-dimensional torus were Markov shifts of a special kind and went on to show that any two Markov shifts of this kind with the same entropy were isomorphic.

Classifying the Bernoulli shifts gives information about them which at first glance one would not expect to get. For example, it was not previously known if the 2-shift had a square root. We can now show it has a square root as follows. Let T be a Bernoulli shift whose entropy is one-half that of the 2-shift. It is easy to see that T^2 will be a Bernoulli shift with the same entropy as the 2-shift. Therefore, the 2-shift has a square root. Similarly, Bernoulli shifts have roots of all orders and have lots of automorphisms that commute with them.

I would like to end by discussing a very general conjecture of Kolmogorov, as yet unproved. Kolmogorov singled out a class of transformations, now called K-automorphisms, and conjectured that they were Bernoulli shifts. (Bernoulli shifts are easily seen to be K-automorphisms.) The question is important because a large number of transformations arising in other contexts can be shown to be K-automorphisms. For example, there are K-automorphisms that are embedded in flows, but we don't know if a Bernoulli shift can be embedded in a flow. We will not give Kolmogorov's original definition but will give an equivalent definition due to Sinai. Let P be a finite partition with sets P_i, and Q a finite partition with sets Q_i. We will say that P is ϵ-independent of Q if there is a collection $\mathcal{E}$ of Q_i, the measure of whose union is greater than $1 - \epsilon$, and if $Q_i \in \mathcal{E}$, then

$$\sum_j |m(P_j \cap Q_i)/m(Q_i) - m(P_j)| < \epsilon.$$

T is a K-automorphism if, given a partition P and ϵ there is an n such that P is ϵ-independent of $\bigvee_{m}^{m+m'} T^i P$ for all $m \geq n$, $m' \geq 0$. If the above holds for a P such that the $T^i P$ generate, then it holds for all P.

We can prove the following. If $T^i P$ generates, and if given ϵ, we can find an n such that $\bigvee_{-m_1}^{0} T^i P$ is ϵ-independent of $\bigvee_{m_2}^{m_2+m_3} T^i P$ for all $m_1 \geq 0$, $m_2 > n$ and $m_3 \geq 0$, then T is a Bernoulli shift.

This theorem is proved in [4], and we get the result about Markov shifts as its corollary. The answer to the following question is not known. If the above property holds for a generator, does it hold for all P? (Equivalently, does the above property hold for all finite partitions if T is a Bernoulli shift?)

REFERENCES

1. D. Ornstein, Bernoulli shifts with the same entropy are isomorphic, to appear.
2. D. Ornstein, Bernoulli shifts with infinite entropy are isomorphic, ibid.
3. D. Ornstein, Factors of Bernoulli shifts are Bernoulli shifts, ibid.
4. D. Ornstein and N. Friedman, Markov Shifts are Bernoulli shifts, ibid.
5. D. Ornstein and M. Smorodinsky, An exposition of Ornstein's isomorphic theorem, ibid.
6. R. L. Adler and B. Weiss, Entropy, a complete metric invariant for automorphisms of the torus, Proc. Nat. Acad. Sci. U.S.A., 57 (1967), pp. 1573-1576.
7. P. Billingsley, Ergodic Theory and Information, New York, John Wiley and Sons, 1965.
8. P. Halmos, Recent progress in ergodic theory, Bull. Amer. Math. Soc., 67 (1961), pp. 70-80.
9. P. Halmos and H. Vaughn, The marriage problem, Amer. J. Math., 72 (1950), pp. 214-215.
10. A. N. Kolmogorov, A new metric invariant of transitive dynamic systems and automorphisms in Lebesgue spaces, Dokl. Akad. Nauk SSSR, 119 (1958), pp. 861-864.
11. A. N. Kolmogorov, On the entropy per unit time as a metric invariant of automorphisms, Dokl. Akad. Nauk SSSR, 124 (1959), pp. 754-755.
12. L. D. Mesalkin, A case of isomorphisms of Bernoulli scheme, Dokl. Akad. Nauk SSSR, 128 (1959), pp. 41-44.
13. I. G. Sinai, A weak isomorphism of transformations with an invariant measure, Dokl. Akad. Nauk SSSR, 147 (1962), pp. 797-800, and Soviet Math. Dokl., 3 (1962), pp. 1725-1729.

ON CERTAIN ANALOGOUS DIFFICULTIES IN THE INVESTIGATION OF FLOWS IN A PROBABILITY SPACE AND OF TRANSFORMATIONS IN AN INFINITE MEASURE SPACE

U. Krengel[1] – *Ohio State University*

SUMMARY. The entropy of a measure preserving flow $\{T_t, t \in \mathbb{R}^1\}$ in a probability space or of a conservative measure preserving transformation T in an infinite, σ-finite measure space can be defined either by reduction to the discrete finite case or by looking at increasing partitions and defining the analog of mean entropy for them. Relations between these definitions are investigated and sufficient condition for their equivalence are given. By proving the existence of a dense system of sets A, for which $\{A, A^c\}$ is a strong generator, it is shown that the usual methods for computing the supremum of the mean entropies (theorem of Kolmogorov-Sinai, continuity of mean entropy, monotonicity of mean entropy) fail for flows and for transformations in infinite measure spaces.

§1. INTRODUCTION. One of the fundamental facts from the entropy theory of measure preserving (m.p.) transformations is the following. The entropy h(t) of a m. p. transformation T in a finite measure space $(\Omega, \mathfrak{S}, \mu)$ is zero if and only if there is no strictly increasing subalgebra $\mathfrak{S}_0 \subset \mathfrak{S}$; i.e. no $\mathfrak{S}_0$ with $T^{-1}\mathfrak{S}_0 \subset \mathfrak{S}_0$ and $T^{-1}\mathfrak{S}_0 \neq$ $\neq \mathfrak{S}_0$ mod μ. (For definitions see below.) Whether or not the corresponding result holds for flows $\{T_t, t \in \mathbb{R}^1\}$ in Ω and for conservative m.p. transformations in an infinite measure space is unknown. The purpose of this talk is to point out a number of close analogies between these two unsolved problems. We obtain only partial results, but we feel that they may help to illuminate the main difficulties. In particular we are led to a study of an alternative concept of entropy for flows. (For m.p. transformations in an infinite measure space analogous results were given by Parry in his book [15]. We give new proofs of these results, not depending on the theory of Lebesgue spaces, and simultaneously generalize them.) The question whether the new entropy $h^*(\{T_t\})$ of flows (Parry's entropy $h^*(T)$) coincides with the entropy $h(\{T_t\})$ as defined by Kolmogorov [9] (resp. with h(T) as defined by the author [10]) is essentially equivalent to the unsolved problem stated above. So far there seems to be no way of computing h^* except by showing that h^* coincides with h for certain flows and transformations. The methods used for the computation of the entropy of a m.p. transformation in a probability space fail in the infinite case and in the case of flows. This is shown by proving the existence of a dense system of strong finite generators.

[1] Research supported by the National Science Foundation, Grant GP-9354.

(Deeper theorems, to be published elsewhere [12], assert the existence of finite strong generators in a given exhaustive sub-σ-algebra. However they are of interest only for entirely different questions. We therefore feel that it is desirable to give here the much shorter proof of the existence theorem in its simplest form. Also, our present theorem 4.1 on the existence of strong generators is not contained in the theorem given in [12], since we prove here the density of the set of strong generators in the original infinite measure space, whereas in [12] we have to pass to a finite equivalent measure.)

We (and other authors) have repeatedly referred to Scheller's [17] unpublished extension of Abramov's formula for the entropy of induced transformations. We therefore give a sketch of Scheller's proof in an appendix with his kind permission. I would like to thank Dr. Scheller for this permission and Mr. A. Borden and Professor E. Klimko for some helpful remarks.

§2. PRELIMINARIES. Let $(\Omega, \mathcal{E}, \mu)$ be a finite or σ-finite measure space. For any sub-σ-algebra $\mathcal{A}$ of $\mathcal{E}$ we denote by $\mathfrak{P}(\mathcal{A})$ the system of all finite partitions $\xi = \{A_0, A_1, \ldots, A_n\}$ with $A_i \in \mathcal{A}$ and $\mu(A_i) < \infty$ $(i = 1, \ldots, n)$, and by $\mathfrak{P}_\sigma(\mathcal{A})$ the system of all finite or countable partitions $\eta = \{B_1, B_2, \ldots\}$ with $B_i \in \mathcal{A}$ and $\mu(B_i) < \infty$ $(i = 1, 2, \ldots)$. If it is obvious to which $\mathcal{A}$ we refer, we shall use the simpler notation $\mathfrak{P}$ and $\mathfrak{P}_\sigma$.

For finite measures ν and partitions $\xi \in \mathfrak{P}$ the entropy of ξ (under ν) is defined by $H(\xi, \nu) = -\sum_{i=0}^{n} \nu(A_i) \log \nu(A_i)$, where $0 \cdot \log 0 = 0$. For any set $B \in \mathcal{E}$ with $0 < \mu(B) < \infty$, μ_B denotes the normalized restriction of μ to B, i.e. $\mu_B(A) = \mu(B)^{-1} \cdot \mu(A \cap B)$. The conditional entropy of $\xi \in \mathfrak{P}$ given $\eta \in \mathfrak{P}_\sigma$ is defined by:

$$H(\xi | \eta) = \sum_{i=1}^{\infty} \mu(B_i) H(\xi, \mu_{B_i}).$$

As in the case of finite measure spaces $\xi_1 \leqslant \xi_2$ implies $H(\xi_1 | \eta) \leqslant H(\xi_2 | \eta)$ and $\eta_1 \leqslant \eta_2$ implies $H(\xi | \eta_1) \geqslant H(\xi | \eta_2)$ For sub-σ-algebras $\mathcal{A}$ and $\mathcal{B}$ of $\mathcal{E}$ let

$$(2.1) \qquad H(\mathcal{A} | \mathcal{B}) = \sup \{ \inf \{ H(\xi | \eta) : \eta \in \mathfrak{P}_\sigma(\mathcal{B}) \} : \xi \in \mathfrak{P}(\mathcal{A}) \}.$$

We shall need (2.1) only in the case where μ is σ-finite on $\mathcal{B}$ and $\mathcal{A}$ is such that $\mathcal{A}$ has at most one atom of infinite measure, i.e. there is some $A_0 \in \mathcal{A}$, possibly empty, such that the restriction of $\mathcal{A}$ to A_0 is $\{A_0, \emptyset\}$ mod μ and the restriction of μ to $A_0^c \cap \mathcal{A}$ is σ-finite. (A^c denotes the complement of A). Such σ-algebras $\mathcal{B}$ will be called σ-finite and such σ-algebras $\mathcal{A}$ will be called σ_0-finite. In the case of Lebesgue spaces, Parry [15] has given an alternative but equivalent definition of $H(\mathcal{A} | \mathcal{B})$ in terms of canonical measures.

We identify sets and σ-algebras coinciding mod μ, and also partitions ξ, η etc. with the σ-algebras they generate. $H(\mathcal{A} | \mathcal{B})$ is zero if and only if $\mathcal{A}$ is a subalgebra of $\mathcal{B}$.

A measure preserving (m.p.) transformation $T: \Omega \to \Omega$ will be called an *endomorphism;* if it is invertible (i.e. 1-1, onto, T^{-1} = endomorphism) it is called an *automorphism* of $(\Omega, \mathcal{E}, \mu)$.

T shall always denote an endomorphism of Ω. $\{T_t\}$ denotes a m.p. measurable flow, i.e. a group of automorphisms T_t such that $T_{t+s} = T_t T_s$ $(t, s \in \mathbb{R}^1)$ and such that the mapping $\Omega \times \mathbb{R}^1 \ni (\omega, t) \to T_t\omega \in \Omega$ is measurable. $\xi(k, \ell) = \xi(k, \ell; T)$ denotes the partition or σ-algebra generated by $T^i\xi$ $(-\infty \leqslant k \leqslant i \leqslant \ell \leqslant \infty)$ and $\xi[s, t]$ denotes the partition generated by the partitions $T_u\xi$ $(-\infty \leqslant s \leqslant u \leqslant t \leqslant \infty)$. For $\xi \in \mathfrak{P} \cup \mathfrak{P}_\sigma$ let $h(\xi, T) = H(\xi \mid \xi(-\infty, -1))$ and $h(\xi, \{T_t\}) = H(\xi \mid \xi[-\infty, -1])$, whenever these conditional entropies are well-defined. $h(\xi, T)$ will be called the *mean entropy* of ξ with respect to T.

The entropy of an endomorphism T of a *finite* measure space is:

$$h(T) = \sup\{h(\xi, T) : \xi \in \mathfrak{P}\}. \tag{2.2}$$

In general it is impossible to compute $h(\xi, T)$ for *all* partitions. Fortunately it is sufficient to compute $h(\xi, T)$ for certain sufficiently fine partitions. ξ is called a *generator for an automorphism T* if $\xi(-\infty, +\infty) = \mathcal{E}$; ξ is called a *strong generator for an endomorphism T* if $\xi(-\infty, 0) = \mathcal{E}$. We call ξ a *generator for $\{T_t\}$* if $\xi[-\infty, +\infty] = \mathcal{E}$ and a *strong generator for $\{T_t\}$* if $\xi[-\infty, 0] = \mathcal{E}$. The computation of $h(T)$ is facilitated by the following results.

(2.3) *Theorem of Kolmogorov-Sinai*. If $H(\xi) < \infty$ and ξ is a generator (strong generator) for the automorphism (endomorphism) T of the finite measure space $(\Omega, \mathcal{E}, \mu)$, then $h(T) = h(\xi, T)$.

(2.4) *Monotonicity of mean entropy*. $\xi \geqslant \eta$, $\xi \in \mathfrak{P}$ implies $h(\xi, T) \geqslant h(\eta, T)$.

(2.5) *Continuity of mean entropy*. If $\xi^k = \{A_1^k, \ldots, A_n^k\}$ converges to $\xi = \{A_1, \ldots, A_n\}$ in the sense that $\sum_{i=1}^{n} \mu(A_i^k \,\Delta\, A_i) \to 0$ $(k \to \infty)$, then $h(\xi^k, T) \to h(\xi, T)$.

In fact, sharper statements than (2.3) - (2.5) hold, and they can be combined.

Kolmogorov [9] defined the entropy of a m.p. flow $\{T_t\}$ in a finite measure space by:

$$h(\{T_t\}) = \sup\{t^{-1} h(T_t) : t > 0\}. \tag{2.6}$$

Abramov [2] proved the equation

$$h(T_t) = |t|\, h(T_1) \tag{2.7}$$

for measurable flows, and Pinsker extended this result to "continuous" m. p. semiflows $\{T_t, t \geqslant 0\}$. (Pinsker's unpublished proof is contained in [11, p. 37].) In particular it follows that

$$h(\{T_t\}) = h(T_1), \tag{2.8}$$

and we may take (2.8) as the definition of the entropy of a flow.

An endomorphism T of an infinite measure space $(\Omega, \mathcal{E}, \mu)$ is called *conservative* if for all $E \in \mathcal{E}$, $E \subset \bigcup_{k=1}^{\infty} T^{-k}E \mod \mu$. In this case

$$r_E(\omega) = \inf\{k \geq 1: T^k\omega \in E\}$$

is finite a.e. on E. (We shall discard exceptional nullsets like this one without mentioning it in the sequel.) $T_E: \omega \to T^{r_E(\omega)}\omega$ is an endomorphism of $(E, E \cap \mathcal{E}, \mu)$. The author [10] has defined the entropy of a conservative endomorphism T by

$$\text{(2.9)} \qquad h(T) = \sup\{h(T_E): 0 < \mu(E) < \infty\}.$$

It can be shown using a theorem of Abramov [1] and its extension by Scheller [17], (see section 5), that $h(T) = h(T_E)$ for every *sweep-out set* E, i.e. for every $E \in \mathcal{E}$ such that $\Omega = \bigcup_{k=0}^{\infty} T^{-k}E$.

Recently Parry [15] suggested that the entropy of an endomorphism T be defined by:

$$\text{(2.10)} \qquad h^*(T) = \sup\{H(\mathcal{E}_0 \mid T^{-1}\mathcal{E}_0): \mathcal{E}_0 \in \Phi\},$$

where Φ is the system of all invariant, σ-finite subalgebras $\mathcal{E}_0$ of $\mathcal{E}$, i.e. the system of all σ-algebras $\mathcal{E}_0 \subset \mathcal{E}$ with $T^{-1}\mathcal{E}_0 \subset \mathcal{E}_0$ such that μ is σ-finite on $\mathcal{E}_0$. For endomorphisms T of a finite measure space one has $h(T) = h^*(T)$. (Using concepts from [8], E. Klimko [6] has independently suggested an alternative definition of entropy, equivalent to Parry's. (Parry's book is based on lectures given during the autumn of 1966 at Yale University, but all results on infinite measure spaces have been added later.))

In a manner similar to (2.10) we define the entropy $h^*(\{T_t\})$ of a m.p. flow $\{T_t\}$ in a finite measure space $(\Omega, \mathcal{E}, \mu)$ by

$$\text{(2.11)} \qquad h^*(\{T_t\}) = \sup\{H(\mathcal{E}_0 \mid T_1^{-1}\mathcal{E}_0): \mathcal{E}_0 \in \Psi\},$$

where Ψ is the system of all invariant subalgebras $\mathcal{E}_0$ of $\mathcal{E}$, i.e. the system of all σ-algebras $\mathcal{E}_0 \subset \mathcal{E}$ with $T_t^{-1}\mathcal{E}_0 \subset \mathcal{E}_0$ for all $t \geq 0$. Of course we could easily extend this definition to the σ-finite case, but we want to emphasize that the same difficulties which we encounter for a fixed T in infinite measure spaces appear already in finite measure spaces when we deal with flows. Therefore, whenever we speak of flows, they will be m.p. flows in a finite measure space.

The questions stated in the introduction can now be reformulated in terms of h and h^*. Is $h(T) = 0$ if $h^*(T) = 0$? Is $h(\{T_t\}) = 0$ if $h^*(\{T_t\}) = 0$? (The converse implications are easy.) More generally we ask: is $h = h^*$?

§ 3. COMPARISON OF h AND h*. h(T) is defined only for conservative endomorphisms, but this restriction is not very serious. First, to quote E. Hopf [5, p. 35], "The conservative part of Ω is the vital part as far as ergodic theory is concerned"; second, $h^*(T)$ does not distinguish between non-conservative automorphisms. It is easy to see that $h^*(T) = \infty$ holds for non-conservative automorphisms -- at least if the measure space is non-atomic. We shall see in a moment that $h^*(T)$ has the form

$$(3.1) \qquad h^*(T) = \sup\{h(\xi, T): \xi \in \mathfrak{P}\}$$

if we remove the difficulty that $h(\xi, T) = H(\xi \mid \xi(-\infty, -1))$ need not be defined since μ need not be σ-finite on $\xi(-\infty, -1)$. Let $\xi = \{A_0, A_1, \ldots, A_n\}$, $\mu(A_i) < \infty$ $(i = 1, \ldots, n)$. μ is σ-finite on $\xi(-\infty, -1)$ when restricted to $\bigcup_{k=0}^{\infty} T^{-k} A_0^c$. $h(\xi, T)$ is defined to be $H(\xi \mid \xi(-\infty, -1))$ when computed on this union. More generally: for countable partitions $\xi = \{A_0, A_1, \ldots\}$, $\eta = \{B_0, B_1, \ldots\}$ with $\mu(B_i) < \infty$ $(i = 1, 2, \ldots)$ and $A_0 \supset B_0$, let $H(\xi \mid \eta) = \sum_{i=1}^{\infty} \mu(B_i) H(\xi, \mu_{B_i})$. We "forget" the set B_0, which may have infinite measure, since it is not subdivided by ξ. In the same way we get an obvious extension of the definition of $H(\mathcal{A} \mid \mathcal{B})$. To prove (3.1) let r be a real number with $r < h^*(T)$. There is an increasing, σ-finite $\mathfrak{E}_0 \subset \mathfrak{E}$ with $H(\mathfrak{E}_0 \mid T^{-1}\mathfrak{E}_0) > r$. Hence there is an $\mathfrak{E}_0$-measurable $\xi \in \mathfrak{P}$ with $H(\xi \mid T^{-1}\mathfrak{E}_0) > r$. It follows from $\xi(-\infty, -1) \subset \subset T^{-1}\mathfrak{E}_0$ that $H(\xi \mid \xi(-\infty, -1)) > r$. Since r was arbitrary we obtain $\leqslant$ in (3.1). On the other hand if $\xi = \{A_0, \ldots, A_n\} \in \mathfrak{P}$ is given, define $\mathfrak{E}_0$ such that it coincides with $\xi(-\infty, 0)$ on the invariant set $\bigcup_{k=0}^{\infty} T^{-k} A^c$ and with $\mathfrak{E}$ on the complement. Then $H(\mathfrak{E}_0 \mid T^{-1}\mathfrak{E}_0) = h(\xi, T)$ and $\geqslant$ follows in (3.1). Similarly we obtain for m.p. flows $\{T_t\}$ in a finite measure space

$$(3.2) \qquad h^*(\{T_t\}) = \sup\{h(\xi, \{T_t\}): \xi \in \mathfrak{P}\}.$$

3.1 PROPOSITION. If T is a conservative endomorphism of a σ-finite measure space $(\Omega, \mathfrak{E}, \mu)$, then $h(T) \geqslant h^*(T)$. If $\{T_t\}$ is a measurable m.p. flow in a finite measure space, then $h(\{T_t\}) \geqslant h^*(\{T_t\})$.

Proof. The second statement follows from $\xi(-\infty, -1, T_1) \subset \xi[-\infty, -1]$ and the monotonicity of $H(\mathcal{A} \mid \mathcal{B})$.

To prove the first statement let $r < h^*(T)$. There is a partition $\xi = \{A_0, A_1, \ldots, A_k\} \in \mathfrak{P}$ with $r < H(\xi \mid \xi(-\infty, -1))$. Let $A = \bigcup_{i=1}^{k} A_i$, $B^n = \bigcup_{i=1}^{n} T^{-i} A$ and $B^\infty = \bigcup_{i=1}^{\infty} T^{-i} A$. For n large enough

$$(3.3) \qquad H(\xi \cap B^n \mid \xi(-\infty, -1) \cap B^n) > r,$$

if this conditional entropy is computed on the smaller measure space $(B^n, \mathfrak{E} \cap B^n, \mu)$. (Note that $B^n \in \xi(-\infty, -1)$!) Fix n so that (3.3) holds, and let $B = B^n$. The main step in the proof consists in verifying

$$(3.4) \qquad (\xi \cap B)(-\infty, -1, T_B) \subset B \cap \xi(-\infty, -1, T).$$

This is a little laborious, but straightforward. From (3.3) and (3.4) it follows that $H(\xi \cap B \mid (\xi \cap B)(-\infty, -1, T_B)) > r$. It follows that $h(T_B) > r$. Since $h(T) \geqslant h(T_B)$ and r was arbitrary with $r < h^*(T)$, we have $h(T) \geqslant h^*(T)$.

Parry [15] has proved $h(T) = h^*(T)$ under the additional assumptions that T is invertible and $(\Omega, \mathcal{E}, \mu)$ is a Lebesgue space. He also has a proof of the next proposition which depends heavily on the theory of Lebesgue spaces.

If T is a conservative endomorphism, ρ_E denotes the ***return partition*** for the set $E \in \mathcal{E}$, i.e. the partition of E into the sets $R_k(E) = \{\omega \in E: r_E(\omega) = k\}$ $(k = 1, 2, 3, \ldots)$. In [10] T was called ***quasifinite*** if there exists at least one sweep-out set $E \in \mathcal{E}$ with $\mu(E) < \infty$ such that $H(\rho_E, \mu_E) < \infty$. Here we shall use the term endomorphism of ***type Z*** instead of the term "quasifinite" endomorphism. (In [10], I expected that such endomorphisms would have basically the same structure as endomorphisms of finite measure spaces. The results of this paper show that this is not so).

3.2. PROPOSITION. Let T be a conservative automorphism of type Z of the σ-finite measure space $(\Omega, \mathcal{E}, \mu)$. Then $h(T) = h^*(T)$.

Proof. Because of Proposition 3.1 it remains to show $h \leqslant h^*$. Let E be a sweep-out set of finite measure with $H(\rho_E, \mu_E) < \infty$. As $h(T)$ and $h^*(T)$ both depend linearly on the invariant measure μ defined on Ω, we may assume $\mu(E) = 1$. Let ξ be a partition of E with $H(\xi, \mu_E) < \infty$. It is sufficient to find a countable partition ζ of Ω such that μ is σ-finite on $\zeta(-\infty, -1, T)$ and

$$(3.5) \qquad H(\zeta \mid \zeta(-\infty, -1, T)) \geqslant H(\xi \mid \xi(-\infty, -1, T_E)) = h(\xi, T_E).$$

Observe that the entropy on the left-hand side is computed on Ω and the entropy on the right-hand side on E! If ξ is replaced by a finer partition ξ^1 with $H(\xi^1, \mu_E) < < \infty$, then $h(\xi^1, T_E) \geqslant h(\xi, T_E)$. We may therefore assume that

$$(3.6) \qquad \rho_E \subset \xi \text{ and } T_E \rho_E \subset \xi,$$

by refining ξ if necessary. Now let ζ be the partition of Ω which is the refinement of the partition $\{TE, TE^c\}$ and the partition consisting of the set E^c and the sets of ξ. Using (3.6) one can show in a straightforward way that

$$(3.7) \qquad \xi(-\infty, -1, T) \cap E \subset \xi(-\infty, -1, T_E).$$

It follows that

$$H(\zeta \mid \zeta(-\infty, -1, T)) \geqslant H(\xi \mid \xi(-\infty, -1, T_E))$$

because E belongs to $\zeta(-\infty, -1, T)$ and the conditional entropy on the left-hand side is greater than or equal to the corresponding entropy computed on the restriction to E.

Next we shall prove the analogue of Proposition 3.2 for m.p. flows. For this we have to recall the concept of a flow under a function. Let $(\Sigma, \mathfrak{F}, \nu)$ be a finite measure space and f a ν-integrable function on Σ, bounded below by some $c > 0$. Let S be an automorphism of Σ. A new measure space and a flow are defined as follows: $\Omega = \{(\sigma, x) \in \Sigma \times \mathbb{R}^1 : 0 \leqslant x < f(\sigma)\}$, μ is the completion of the restriction of the product measure $\nu \times \lambda$ to Ω, where λ is Lebesgue measure in $\mathbb{R}^1$. The flow $\{T_t\}$ is defined for $0 \leqslant t < c$ by: $T_t(\sigma, x) = (\sigma, x + t)$ if $x + t < f(\sigma)$, $T_t(\sigma, x) = (S\sigma, x + t - f(\sigma))$ if $x + t \geqslant f(\sigma)$. Every ergodic m.p. measurable flow is isomorphic mod sets of measure zero to such a flow [3]. A flow $\{T_t\}$ will be called a *flow of type Z* if there is an isomorphic representation such that f takes only countably many values $a_1, a_2, a_3, \ldots,$ and the decomposition ρ of Σ into the sets $\{\sigma \in \Sigma: f(\sigma) = a_i\}$ has finite entropy. We do not know whether every ergodic m.p. flow is of type Z. (Similarly it is quite difficult to show that there are ergodic automorphisms which are not of type Z. D. Ornstein recently informed me by phone about a sketch of a construction of such an automorphism.)

3.3. PROPOSITION. If $\{T_t\}$ is a measurable m.p. flow of type Z in the finite measure space $(\Omega, \mathcal{E}, \mu)$, then $h(\{T_t\}) = h^*(\{T_t\})$.

Proof. For $a > 0$ define the flow $\{T_t^{(a)}\}$ by $T_t^{(a)} = T_{at}$. Since $h^*(\{T_t\})$ and $h(\{T_t\})$ depend linearly on the invariant measure, we may assume $\mu(\Omega) = 1$ for a moment. $\mathcal{E}_0 \subset \mathcal{E}$ is $\{T_t\}$-invariant if and only if it is $\{T_t^{(a)}\}$-invariant. For any invariant $\mathcal{E}_0 \subset \mathcal{E}$ the function $g(a) = H(\mathcal{E}_0 \mid T_a^{-1}\mathcal{E}_0)$ is monotone increasing and satisfies

$$\begin{aligned} g(a_1 + a_2) &= H(\mathcal{E}_0 \mid T_{a_1+a_2}^{-1}\mathcal{E}_0) \\ &= H(T_{a_1}^{-1}\mathcal{E}_0 \mid T_{a_1+a_2}^{-1}\mathcal{E}_0) + H(\mathcal{E}_0 \mid T_{a_1}^{-1}\mathcal{E}_0) \\ &= g(a_1) + g(a_2). \end{aligned}$$

This implies $h^*(\{T_t^{(a)}\}) = ah^*(\{T_t^{(a)}\})$. Similarly (2.7) implies $h(\{T_t^{(a)}\}) = ah(\{T_t\})$. It is therefore sufficient to prove the desired equation for some $\{T_t^{(a)}\}$ instead of $\{T_t\}$. $\{T_t^{(a)}\}$ has a representation as a flow under a function which is obtained from the representation of $\{T_t\}$ by replacing ν by $a\nu$, f by $a^{-1}f$ and c by $a^{-1}c$. We may therefore restrict our attention to flows for which the bounding constant c is larget than 2. (The index a will be dropped now.)

For the second part of the argument we normalize μ such that $\nu(\Sigma) = 1$. It is known that $h(\{T_t\}) = h(S)$, (see [2] or [11]). In view of Proposition 3.1 it suffices to prove $h^*(\{T_t\}) \geq h(S)$. Let $r < h(S)$ and ξ a countable partition of Σ such that $H(\xi, \eta) < \infty$ and

$$(3.8) \qquad h(\xi, S) = H(\xi \mid \xi(-\infty, -1, S)) > r.$$

We may assume that ξ is finer than the partitions $\rho = \{\{f = a_1\}, \{f = a_2\}, \ldots\}$ and $S\rho$ by refining ξ if necessary. Let $\mathcal{E}_0$ be the sub-σ-algebra of $\mathcal{E}$ which is generated by the sets $R(x_1, x_2) = \{(\sigma, x) \in \Omega: x_1 \leq x < x_2\}$ and by the sets $B^* = \{(\sigma, x) \in \Omega: \sigma \in B\}$ for $B \in \xi(-\infty, 0, S)$. $\mathcal{E}_0$ is invariant because for $0 < t \leq 1$, $T_t^{-1}\mathcal{E}_0$ is the σ-algebra $\mathcal{R}$ generated by the sets $R(x_1, x_2)$, the sets $B^* \cap R(0, t)$ with $B \in \xi(-\infty, -1, S)$, and $B^* \cap R(t, t+s)$ with $B \in \xi(-\infty, 0, S)$. In particular $T_1^{-1}\mathcal{E}_0$ coincides with $\mathcal{E}_0$ on $\{(\sigma, x): x \geq 1\}$. On $\Omega_{0,1} = \{(\sigma, x): 0 \leq x < 1\}$, $\mathcal{E}_0$ is the product of the independent σ-algebras $\xi(-\infty, 0, S) \times [0, 1]$ and $\mathcal{R} \cap \Omega_{0,1}$, and $T_1^{-1}\mathcal{E}_0$ is the product of the independent σ-algebras $\xi(-\infty, -1, S) \times [0, 1]$ and $\mathcal{R} \cap \Omega_{0,1}$. Because of this independence we have

$$H(\mathcal{E}_0 \mid T_1^{-1}\mathcal{E}_0) = H(\xi(-\infty, 0, S) \mid \xi(-\infty, -1, S)) = h(\xi, S).$$

Hence $h^*(\{T_t\}) > r$, and since $r < h(\{T_t\})$ was arbitrary, the desired inequality follows.

Very often it will be easier to prove $h = h^*$ directly instead of using Propositions 3.2 and 3.3. Let for example I be the countable state space and $\{P_{ik}; i, k \in I\}$ the transition matrix of a null-recurrent Markov chain with invariant measure $\{\pi_i, i \in I\}$. Let $\Omega_j = I$, $\Omega = \prod_{j=-\infty}^{\infty} \Omega_j$ and $\mathcal{E}$ the usual product σ-algebra. For any $\omega = (\ldots, \omega_{-1}, \omega_0, \omega_1, \ldots) \in \Omega$ let $(\omega)_i$ be its i-th coordinate ω_i. Let μ be the unique measure for which

$$\mu(\{\omega: (\omega)_k = i_k, (\omega)_{k+1} = i_{k+1}, \ldots, (\omega)_{k+n} = i_{k+n}\}) =$$

$$= \pi_{i_k} \cdot P_{i_k i_{k+1}} \cdots P_{i_{k+n-1} i_{k+n}}.$$

The shift T in Ω defined by $(T\omega)_i = (\omega)_{i+1}$ is a conservative automorphism of Ω. It has been shown in [10, p. 167 and p. 179] that $h(T) = h(T^{-1}) = -\Sigma_i \pi_i \Sigma_k P_{ik} \log P_{ik}$. Now let ξ be the partition of Ω into the sets $A_i = \{\omega: (\omega)_0 = i\}$ $(i \in I)$. The Markov property implies

$$H(\xi \mid \xi(-\infty, -1, T^{-1})) = H(\xi \mid T\xi) = \Sigma_i \Sigma_k \pi_i P_{ik} \log P_{ik}.$$

Since $h^*(T^{-1}) \leq h(T^{-1})$, it follows that $h^*(T^{-1}) = h(T^{-1})$. (It is not quite clear whether $h^*(T) = h^*(T^{-1})$ holds for all conservative automorphisms.) Similar considerations can be carried out for Markov-flows: start out with a standard continuous time parameter Markov chain $(P_{ik}(t))$ with finitely many states, all states being recurrent. Except in the purely deterministic case ($P_{ii}(t) = 1$ for all i) when $h(\{T_t\}) = 0$

one obtains $h^*(\{T_t^{-1}\}) = \infty$, and hence $h^*(\{T_t^{-1}\}) = h(\{T_t^{-1}\}) = \infty$.

If T is an automorphism of a finite measure space, then $H(\rho_E) < \infty$ holds for all sets $E \in \mathcal{E}$. The following proposition shows that in the ergodic infinite case there always exist sets E with $\mu(E) < \infty$ such that $H(\rho_E) = \infty$. In fact there are always sets which behave "arbitrarily badly". In particular this implies that all sets $E \in \mathcal{E}$ with finite measure must be considered before the hypothesis that T is of type Z is rejected.

3.4. PROPOSITION. Let T be an ergodic automorphism of a σ-finite, infinite, nonatomic measure space $(\Omega, \mathcal{E}, \mu)$ and let $r_1, r_2, r_3, \ldots$ be a sequence of positive real numbers with $r = \Sigma\, r_i < \infty$. For any $\epsilon > 0$ there exists a set $E \in \mathcal{E}$ with $\mu(E) < r + \epsilon$ and $\mu(R_k(E)) \geqslant r_k$ $(k = 1, 2, \ldots)$.

For the proof we require a lemma which resembles a well-known theorem of Rokhlin on aperiodic transformations (lemma 2.3 in [10]), and is in fact derived from it.

A conservative endomorphism T of $(\Omega, \mathcal{E}, \mu)$ is called *aperiodic* if there is no $n \geqslant 1$ such that for some $A \in \mathcal{E}$ with $\mu(A) > 0$ and all measurable $B \subset A$ the sets B and $T^{-n}B$ are equal mod μ.

3.5. LEMMA. If T is an aperiodic conservative endomorphism of an infinite, σ-finite measure space, then for any $a > 0$ and any n there exists a set $A \in \mathcal{E}$ with $\mu(A) = a$ such that the sets $A, T^{-1}A, \ldots, T^{-n}A$ are disjoint.

Proof. Let $E \in \mathcal{E}$ be such that $\mu(E) > (n+1)a$. By Lemma 2.4 [10] T_E is aperiodic. By Rokhlin's theorem (Lemma 2.3 [10]) there is a set $A \subset E$ with $\mu(A) \geqslant a$ such that $A, T_E^{-1}A, \ldots, T_E^{-n}A$ are disjoint. μ must be nonatomic. Making A a little smaller we get $\mu(A) = a$. The disjointness of the sets $T_E^{-k}A$, $k = 0, \ldots, n$, implies that of the sets $T^{-k}A$, $k = 0, \ldots, n$.

Proof of Proposition 3.4. Pick a natural number n_1 and a set $A_1 \in \mathcal{E}$ such that

$$r_1 < n_1\mu(A_1) < (n_1+1)\mu(A_1) < (r_1 + \epsilon/2) \tag{3.9}$$

and the sets $A_1, T^{-1}A_1, \ldots, T^{-n_1}A_1$ are disjoint. Let $A_{11} = A_1$, $B_{11} = \bigcup_{j=0}^{n_1} T^{-j}A_{11}$. When A_{ik}, B_{ik} and n_i have been defined for $i \leqslant k < m \geqslant 2$ pick $n_m > n_{m-1}$ and A_m such that

$$r_m < n_m\mu(A_m) < (n_m+1)\mu(A_m) < r_m + 2^{-m}\epsilon, \tag{3.10}$$

and the sets $A_m, T^{-1}A_m, \ldots, T^{-m\cdot n_m}A_m$ are disjoint. Let $B_m = \bigcup_{j=0}^{m\cdot n_m} T^{-j}A_m$. Since T is ergodic and the sets B_m and $\bigcup_{i=1}^{m-1} B_{i,m-1}$ have finite measure, we get

$$\liminf_{\ell\to\infty} \mu\Big(\bigcup_{i=1}^{m-1} B_{i,m-1} \cap T^{-\ell}\Big(\bigcup_{h=0}^{m\cdot n_m} T^h B_m\Big)\Big) = 0. \tag{3.11}$$

Now let $A_{i,m} = A_{i,m-1} \cap T^{-\ell}(\bigcup_{h=0}^{m\cdot n_m} T^h B_m))^c$. For arbitrarily large values of ℓ we still have

$$(3.12) \qquad r_i < n_i \mu(A_{i,m}) < (n_i + 1)\mu(A_{i,m}) < r_i + 2^{-i}\epsilon$$

for $i = 1, \ldots, m-1$. Such an ℓ is fixed and we set $A_{m,m} = T^{-\ell} A_m$, $B_{m,m} = \bigcup_{j=0}^{m\cdot n_m} T^{-j} A_{m,m}$. By our construction the sets $B_{i,m} = \bigcup_{j=0}^{i\cdot n_i} T^{-j} A_{i,m}$ are disjoint ($i = 1, \ldots, m$). This completes step number m. For fixed i the sets $A_{i,m}$ decrease as $m \to \infty$ to a set $A_{i,\infty}$ for which (3.12) holds with $m = \infty$ and $\leqslant$ instead of $<$. Now let $C_i = \bigcup_{j=0}^{n_i} T^{-ij} A_{i,\infty}$ and $E = \bigcup_{i=1}^{\infty} C_i$. Since the sets C_i are disjoint from the sets $\bigcup_{j=0}^{k\cdot n_k} T^{-j} A_{k,\infty}$ for $k \neq i$, it follows that $R_i(E) \supset \bigcup T^{-ij} A_{i,\infty}$. In particular $r_i \leqslant R_i(E)$. On the other hand, $\mu(E) < r + \epsilon$ follows immediately from our definitions.

§ 4. EXISTENCE OF STRONG FINITE GENERATORS. The question of existence of finite generators for ergodic automorphisms of a probability space is closely connected with entropy theory. Transformations with infinite entropy do not have finite generators. Ergodic automorphisms T with finite entropy $h(T)$ have finite generators $\xi = \{A_1, \ldots, A_n\}$, and the minimal size n is between $e^{h(T)}$ and $e^{h(T)} + 1$. These results, due to Krieger [13], are deep and require for their proof several powerful tools from ergodic theory. It is therefore perhaps surprising that the existence of finite generators for ergodic automorphisms of an infinite measure space is completely unrelated to entropy theory and can be proved in a comparatively elementary way. In fact, it turns out that even finite strong generators always exist. (Moreover the condition of ergodicity, which is easily seen to be necessary in Krieger's theorem, is not needed for the present result.)

4.1. THEOREM. Let T be an ergodic conservative automorphism of the σ-finite, infinite measure space $(\Omega, \mathcal{E}, \mu)$. Let $\mathcal{E}$ be countably generated mod μ. Then the system of sets $A \in \mathcal{E}$ which are strong generators (i.e. for which $A, T^{-1}A, T^{-2}A, \ldots$ generate $\mathcal{E}$ mod μ) is dense in the set $\mathcal{D}$ of sets of finite measure.

Proof. Let $F_1, F_2, F_3, \ldots$ be a sequence of measurable sets generating $\mathcal{E}$. We may assume that all sets F_i have finite measure and occur infinitely often in the sequence. Let $\epsilon > 0$ and $G \in \mathcal{D}$ be given. It is sufficient to show that there is a set $A \in \mathcal{D}$ with $\mu(A \,\Delta\, G) < \epsilon$ such that the σ-algebra $\delta(A)$ generated by $A, T^{-1}A, T^{-2}A, \ldots$ contains sets F_i^* with $\mu(F_i \,\Delta\, F_i^*) < i^{-1}$. Since T is conservative and ergodic, the measure space $(\Omega, \mathcal{E}, \mu)$ must be nonatomic. Let $G_1, G_2, \ldots \in \mathcal{D}$ be disjoint sets of positive measure with $\Sigma\, \mu(G_k) < \epsilon/2$. We shall first find a countable strong generator with the property that it remains a countable strong generator under sufficiently small changes of the sets. Since T is ergodic, $\bigcup_{j=0}^{\infty} T^{-j} G_k = \Omega$ for all k, and there exist numbers n_k such that $\mu(F_k \setminus \bigcup_{j=0}^{n_k} T^{-j} G_k) < 1/2k$. Let η_k be the finite partition of G_k generated by the sets $T^j(F_k \cap T^{-j}G_k)$ $(j = 0, \ldots, n_k)$, and let $\{E_{k,1}, \ldots, E_{k,m_k}\}$ be the sets of this partition. The set $F_k^1 = F_k \cap \bigcup_{j=0}^{n_k} T^{-j}G_k$ is a finite union of sets of the form $T^{-j}E_{k,i}$ with $0 \leqslant j \leqslant n_k$ and satisfies $\mu(F_k \Delta F_k^1) > 1/2k$. Therefore there exist positive numbers $\epsilon_{k,i}$ $(i = 1, \ldots, m_k)$ such that for any system $E_{k,i}^* \in \mathcal{E}$ with $\mu(E_{k,i} \Delta E_{k,i}^*) < \epsilon_{k,i}$ the set F_k^* which is formed in the same way from the sets $T^{-j}E_{k,i}^*$ as F_k^1 was formed from the sets $T^{-j}E_{k,i}$ satisfies $\mu(F_k^1 \Delta F_k^*) < 1/2k$. It is therefore now sufficient to find a set $A \in \mathcal{E}$ with $\mu(A \Delta G) < \epsilon$ such that $\delta(A)$ contains for all $k \geqslant 1$ and $1 \leqslant i \leqslant m_k$ sets $E_{k,i}^*$ with $\mu(E_{k,i} \Delta E_{k,i}^*) < \epsilon_{k,i}$.

Let $D_1, D_2, \ldots$ be a new enumeration of the sets $E_{k,i}$, and let $\delta_j = \epsilon_{k,i}$ for the index j with $D_j = E_{k,i}$. The sets D_j are disjoint and the measure of their union is less than $\epsilon/2$. Let $D^0 = D = G \cup \bigcup_{j=1}^{\infty} D_j$. Since T is ergodic it follows that

$$(4.1) \qquad \frac{1}{n}\sum_{k=0}^{n-1} \mu(B_1 \cap T^{-k}B_2) \to 0$$

for any two sets $B_1, B_2 \in \mathcal{B}$. Therefore there exists an arbitrarily large k_1 such that

$$(4.2) \qquad \mu(D \cap T^{-k_1}D) < \delta_1/8 \text{ and } \mu(D \cap T^{-2k_1}D) < \delta_1/8.$$

Put $D^1 = D \cup T^{k_1}D_1$; then $D^1 \cap T^{-k_1}D^1 = (D \cap T^{-k_1}D) \cup D_1 \cup (T^{k_1}D_1 \cap T^{-k_1}D) \cup \cup (T^{k_1}D_1 \cap D_1)$ differs from D_1 by a set of μ-measure less than $3\delta_1/8$, since $D_1 \subset \subset D$. At the end of the r-th step of the construction, numbers $k_1 < k_2 < \ldots < k_r$ have been defined such that the sets $D^\ell = D \cup T^{k_1}D_1 \cup T^{k_2}D_2 \cup \ldots \cup T^{k_\ell}D_\ell$ $(\ell = 1, \ldots, r)$ have the properties

$$(4.3) \qquad \mu(D_i \Delta (D^\ell \cap T^{-k_i}D^\ell)) < 3\delta_i \cdot 2^{-2}(2^{-i} + 2^{-(i+1)} + \ldots + 2^{-\ell})$$

for all $i \leqslant \ell \leqslant r$. Let $C^r = \bigcup_{j=-k_r}^{k_r} T^jD^r$. Find $k_{r+1} > k_r$ such that

$$(4.4) \qquad \max\ \mu(C^r \cap T^{-2k_{r+1}}C^r), \mu(C^r \cap T^{-k_{r+1}}C^r) < < 2^{-r-4}\min\{\delta_1, \ldots, \delta_{r+1}\}.$$

It follows that $D^{r+1} = D^r \cup T^{k_{r+1}}D_{r+1}$ satisfies (4.3) for $\ell = r+1$ and $i \leqslant r+1$. We claim that the set $A = \bigcup_{r=1}^{\infty} D^r = D \cup \bigcup_{r=1}^{\infty} T^{k_r}D_r$ has the desired properties. Obviously we have $\mu(G \Delta A) < \epsilon$. Since (4.3) holds for all $\ell \geqslant i$ we get

$$\mu(D_i \Delta (A \cap T^{-k_i}A)) < \delta_i$$

for all i. If j is the index for which $D_j = E_{k,i}$ the set $A \cap T^{-k_j}A$ can serve as the needed set $E_{k,i}^*$.

It follows from Theorem 4.1 that in infinite measure spaces the statements which correspond to the Kolmogorov-Sinai theorem, and to the assertions of the monotonicity of mean entropy and continuity of mean entropy, fail to hold. This was known in part from the author's paper [10], where an example of an ergodic automorphism T of an infinite measure space was given, for which the entropy $h(T)$ was positive and a strong generator $\xi \in \mathfrak{P}$ existed. It is not easy to see whether T in this example was of type Z. Our theorem shows that there are ergodic automorphisms of an infinite measure space which are of type Z, have positive entropy $h(T) = h^*(T) > 0$ and have strong generators of size 2. Since, in fact, the set of strong generators is dense in $\mathfrak{P}$, the mean entropy $h(\xi, T)$ is a discontinuous function on $\mathfrak{P}$. Parry [15, p. 117] remarks that the concept h^* seems to have some advantage over h because of its freedom from induced automorphisms. Unless a new bound for h^* is discovered, h seems to have some advantage over h^*, too, since it is accessible to computation. Theorem 3.1 can easily be extended to the nonergodic case. We only need assume that T is conservative and that there is no invariant subset $I \in \mathfrak{C}$ $(T^{-1}I = I)$ such that T restricted to I has a finite invariant measure equivalent to μ. This is sufficient for (4.1) and implies the aperidoicity of T. It has been shown in [10, p. 165] that aperiodic conservative automorphisms have a sweep-out set D_1 of arbitrarily small measure. The existence of countably many disjoint sweep-out sets D_i follows by showing that any sweep-out set contains two sweep-out sets by the same argument as in [10] and by repeating this splitting procedure (see, e.g., [4]).

The proof of Theorem 4.1 can be modified so as to apply to nonsingular invertible transformations without finite invariant measure. Extending a result of Rokhlin [16] on m.p. transformations, Helmberg and Simons [4] have shown that nonsingular transformations have countable strong generators, provided they are essentially countable-to-one. We do not know whether m-to-one nonsingular transformations without finite invariant measure always have finite strong generators.

Next we shall see that the analog of Theorem 4.1 for flows can be obtained from the representation theorem of Ambrose [3].

4.2. THEOREM. Let $(\Omega, \mathfrak{C}, \mu)$ be a probability space for which $\mathfrak{C}$ is countably generated mod μ, and let $\{T_t\}$ be a measurable m.p. ergodic flow in Ω. Then the system of sets $A \in \mathfrak{C}$ such that $\xi = \{A, A^c\}$ is a strong generator for $\{T_t\}$ (i.e. a partition for which the σ-algebra $\delta(A)$ generated by the sets $\{T_t^{-1}A, t \geqslant 0\}$ equals $\mathfrak{C}$ mod μ) is dense in $\mathfrak{C}$.

Proof. If $(\Omega, \mathfrak{C}, \mu)$ is not complete, we may pass to its completion without losing the measurability of the flow. We may therefore assume that $(\Omega, \mathfrak{C}, \mu)$ is complete. By the representation theorem of Ambrose we may assume that $\{T_t\}$ is a flow under a function $f > c > 0$ as in Section 3. The σ-algebra $\mathfrak{B}$ of the base space $(\Sigma, \mathfrak{B}, \nu)$ is countably generated mod ν, since the dimension of the Hilbert space $\mathcal{L}_2(\Sigma, \mathfrak{B}, \nu)$ is at most equal to the dimension of $\mathcal{L}_2(\Omega, \mathfrak{C}, \mu)$. Let $B \in \mathfrak{C}$ and $\epsilon > 0$ be given. We must find a strong generator $\xi = \{A, A^c\}$ with $\mu(A \Delta B) < \epsilon$. For sufficiently large $K > 0$ and sufficiently small $d > 0$ with $d < c$ the set $B_{d,K} = B \cap \{(\sigma, x) \in \in \Omega\colon d \leqslant x \leqslant K\}$ differs from B by a set of μ-measure less than $\epsilon/3$. We may assume

$\nu(\Sigma) = 1$ and $d < \epsilon/3$. Let $b = d/8$. For $0 < \gamma < b$ sufficiently small the set $E\gamma = \{(\sigma, x) \in \Omega: x \in \bigcup_{k=1}^{\infty} \{y \in R^1: kb \leqslant y \leqslant kb + \gamma\}\}$ has μ-measure less than $\epsilon/3$. Let $G_1, G_2, G_3, \ldots \in \mathcal{B}$ generate $\mathcal{B}$ mod ν, and define g on Σ by

$$g(\sigma) = 4b + 2b \cdot \sum_{k=1}^{\infty} 3^{-k} 1_{G_k}.$$

The set

$$A = (B_{d,K} \cap E_{\gamma}^{c}) \cup \{(\sigma, x) \in \Omega: 2b \leqslant x < g(\sigma)\}$$

clearly satisfies $\mu(A \Delta B) < \epsilon$. Pick an integer m with $m^{-1}b < \gamma$. Let $a = m^{-1}b$. The set

$$C = \bigcap_{i=0}^{m} T_a^{-1} A = \{(\sigma, x) \in \Omega: 2b \leqslant x < g(\sigma) - b\}$$

belongs to $\delta(A)$. Thus it suffices to prove $\mathcal{E} = \delta(C)$. $\delta(C)$ contains the sets $\Omega(x, y) = \{(\sigma, x): \sigma \in \Sigma, x \leqslant z < y\}$ for $0 \leqslant x < y \leqslant 2b$, because for $0 \leqslant b < x < y = 2b$ these sets are of the form $T_{2b-x}^{-1} C \cap C^c$, and the remaining sets can be obtained by applying some T_t^{-1} to these sets and by forming unions.

Let $h(\sigma) = g(\sigma) - b$. Next we show that for arbitrary u with $3b < u < 4b$ the set $H_u = \{(\sigma, x): 0 \leqslant x < 2b, h(\sigma) < u\}$ belongs to $\delta(C)$. Choose N such that $v = u + N^{-1}b < 4b$. The set

$$\bigcup_{i=0}^{2N-1} T_{N^{-1}b}^{-i} (\Omega(2b - N^{-1}b, 2b) \cap T_{v-2b}^{-1} C) = H_{u,N}$$

is contained in H_u and contains H_v. As $N \to \infty$ the sets $H_{u,N}$ increase to H_u mod μ, but the σ-algebra generated by the sets $\{\sigma: h(\sigma) < u, 3b < u < 4b\}$ is $\mathcal{B}$. Thus $\delta(C)$ contains all $\mathcal{E}$-measurable subsets of $\Omega(0, 2b)$, and from this it clearly follows that $\delta(C) = \mathcal{E}$.

It is clear that Theorem 4.2 has the same implications for $h(\{T_t\})$ and $h^*(\{T_t\})$ that Theorem 4.1 has for $h(T)$ and $h^*(T)$. Again the ergodicity is not essential for the result and could be replaced by the weaker assumption that the flow is proper, at the cost of a technically somewhat more complicated proof. ($\{T_t\}$ is called *proper* if every $A \in \mathcal{E}$ contains a $B \in \mathcal{E}$ such that for some t $\mu(B \Delta T_t B) > 0$.)

It follows from Theorem 4.2 that every ergodic m.p. flow in a Lebesgue space is isomrophic to a stationary point process. This observation has independently been made by Professor F. Papangelou.

5. APPENDIX: SCHELLER'S EXTENSION OF ABRAMOV'S FORMULA. In this section we shall give a sketch of Scheller's unpublished proof and extension of Abramov's formula for the entropy of induced endomorphisms, because we have referred to it in this and other papers. The proof is more elementary than the original proof of Abramov [1] in that the theorem of McMillan and the theory of Lebesgue spaces is not required. While Abramov dealt with automorphisms only, Scheller proved the the-

orem in the considerably more intricate case of endomorphisms. In the case of automorphisms most of the steps of the present proof are unnecessary and a rather simple proof of Abramov's formula is obtained. (If Ω is a Lebesgue space the case of endomorphisms can also be reduced to the case of automorphisms by a passage to the invertible extension and using (5.5) below.) A more recent elementary proof [14] of Abramov's formula for automorphisms seems to be incorrect. (The author assumes that $h(T) = h(\xi, T)$ for arbitrary countable generators ξ, but for this $H(\xi) < \infty$ is needed.)

5.1. THEOREM. If T is an endomorphism of the probability space $(\Omega, \mathfrak{E}, \mu)$ and $E \in \mathfrak{E}$ a sweep-out set, then $h(T) = h(T_E)$.

Sketch of proof. Again $r_E(\omega)$ denotes $\inf\{k: T^k\omega \in E\}$, $R_k(E) = \{\omega \in E: r_E(\omega) = k\}$, $\rho_E = \{R_1(E), R_2(E), \ldots\}$. Let $\delta_E = \{E, E^c\}$ and $\rho_E^1 = \{E^c, \rho_E\}$. For any partition or σ-algebra ξ, $\xi^{(E)}$ denotes the partition consisting of E^c and the intersections of the sets from ξ with E. $\xi \vee \eta$ denotes the refinement of ξ and η.

(5.1)
$$\text{For } \xi \in Z^{(E)} = \{\xi \in \mathfrak{P}_\sigma : H(\xi) < \infty, \xi^{(E)} = \xi\}:$$
$$h(\xi, T) \geqslant H(\xi \mid \xi(-\infty, -1, T)^{(E)}) \geqslant h(\xi, T) - H(\delta_E).$$

(5.2)
$$\text{For } \xi \in \mathfrak{P}_\sigma \text{ with } \xi \supset \rho_E^1 \text{ and } \xi = \xi^{(E)}:$$
$$\xi(-\infty, -1, T)^{(E)} \subset \xi(-\infty, -1, T_E) \vee \rho_E^1.$$

This implies

(5.3)
$$\text{For } \xi \in Z^{(E)} \text{ with } \xi \supset \rho_E^1:$$
$$h(\xi, T) \geqslant h(\xi, T_E) - H(\rho_E^1).$$

Since $h(T_E)$ is obtained by taking the supremum of the values of $h(\xi, T_E)$ for the ξ admitted in (5.3) it follows that:

(5.4)
$$h(T) \geqslant h(T_E) - H(\rho_E^1).$$

Next we show

(5.5)
$$h(T_E) = h(T_{T^{-1}E}).$$

For $\xi \in Z^{(E)}$ we have $T^{-1}\xi \in Z^{(T^{-1}E)}$. For $\eta = \{(T^{-1}E)^c, A_1, A_2, \ldots\} \in Z^{(T^{-1}E)}$ the partition $\eta' = \{E, A_1', A_2', \ldots\}$ with $A_k' = \sum_{i=1}^{\infty} R_i(E) \cap T^{-(i-1)}A_k$ satisfies $T_E^{-1}\eta' = T_{T^{-1}E}^{-1}\eta$ and therefore $h(\eta', T_E) = h(\eta, T_{T^{-1}E})$. (5.5) follows by a passage to supremums.

If $E \in \mathcal{E}$ is a sweep-out set, then

$$(5.6) \qquad h(T) = \sup\{\sup\{h(\xi, T: \xi \in Z^{(T^{-n}E)}\}: n \geq 1\}.$$

To show (5.6) it suffices to prove that for any finite $\xi = \{A_1, \ldots, A_k\}$ and $\epsilon > 0$ there is an n and a $\eta \in Z^{(T^{-n}E)}$ with $h(\xi, T) < h(\eta, T) + \epsilon$. Pick $\delta > 0$ such that $|x_i - y_i| < \delta$ $(i = 1, \ldots, 2k+1)$ implies $|\sum_{i=1}^{2k+1} x_i \log x_i - \sum_{i=1}^{2k+1} y_i \log y_i| < \epsilon/2$. For n sufficiently large $\mu(\bigcup_{\ell=0}^{n} T^{-\ell}E) > 1 - \delta$. The partition $\xi' = \{A_0', A_1', \ldots, A_k'\}$ with $A_0' = (\bigcup_{\ell=0}^{n} T^{-\ell}E)^c$, $A_i' = A_i \cap \bigcup_{\ell=0}^{n} T^{-\ell}E$ $(i = 1, \ldots, k)$ satisfies $h(\xi, T) \leq$ $\leq h(\xi', T) + \epsilon$. Define a partition of Ω by

$$E_t = T^{-t}E \cap \bigcap_{i=0}^{t-1} T^{-i}E^c \ (0 \leq t < n), \ E_n = \bigcap_{i=0}^{n} T^{-i}E^c,$$

and a partition ξ'' with atoms $A_k^t = A_k' \cap E_t$. Let η be generated by $T^{-n}E$ and the sets $T^{-(n-t)}A_k$ $(t < n)$. η belongs to $Z^{(T^{-n}E)}$. Since $T^{-n}\xi'' \subset \eta(-\infty, -1, T)$ and $\xi' \subset \xi''$ it follows that $h(\xi, T) \leq h(\xi', T) + \epsilon \leq h(\eta, T) + \epsilon$.

$$(5.7) \qquad \xi(-\infty, -1, T_E) \subset \xi(-\infty, -1, T)^{(E)}.$$

By (5.1) we have

$$H(\xi \mid \xi(-\infty, -1, T_E)) \geq H(\xi \mid \xi(-\infty, -1, T)^{(E)}) \geq h(\xi, T) - H(\delta_E).$$

If we apply (5.5) and (5.6) with $T^{-n}E$ instead of E we get

$$\begin{aligned} h(T) &\leq \sup_n \{\sup\{h(\xi, T_{T^{-n}E}) + H(\delta_{T^{-n}E}): \xi \in Z^{(T^{-n}E)}\}\} \\ &\leq \sup_n \{h(T_{T^{-n}E}) + H(\delta_E)\}. \end{aligned}$$

Thus

$$(5.8) \qquad h(T) \leq h(T_E) + H(\delta_E).$$

Neveu proved the following inequality by a variational argument. If $(p_n)_{n=0}^{\infty}$ is a sequence of nonnegative numbers such that $\sum_{n=0}^{\infty} p_n = \sum_{n=0}^{\infty} n \cdot p_n = 1$, then

$$(5.9) \qquad \sum_{n=0}^{\infty} p_n \log p_n \geq 2(p_0 \log p_0 + (1 - p_0) \log(1 - p_0)).$$

(5.9) and the recurrence theorem of Kac imply

$$(5.10) \qquad H(\rho_E^1) \leq 2H(\delta_E).$$

For the proof of the theorem we may assume that T is aperiodic. By [10, p. 165], each sweep-out set E contains a sweep-out set F of arbitrarily small measure. Applying the above inequalities to $T_F = (T_E)_F$ we get

$$h(T_F) - 2H(\delta_F) \leqslant h(T) \leqslant h(T_F) + H(\delta_F)$$

and the same inequalities for $h(T_E)$ instead of $h(T)$. Since $H(\delta_F) \to 0$ as $\mu(F) \to 0$ it follows that $h(T) = h(T_E)$.

REFERENCES

1. L. M. Abramov, On the entropy of a derived automorphism, Dokl. Akad. Nauk SSSR, 128 (1959), pp. 647-650 = Amer. Math. Soc. Transl. (2), 49 (1966), pp. 162-166.
2. L. M. Abramov, On the entropy of a flow, Dokl. Akad. Nauk SSSR, 128 (1959), pp. 873-876 = Amer. Math. Soc. Transl. (2), 49 (1966), pp. 167-170.
3. W. Ambrose, Representation of ergodic flows, Ann. of Math., 42 (1941), pp. 723-739.
4. G. Helmberg and F. H. Simons, Aperiodic transformations, Z. Wahrscheinlichkeitstheorie verw. Geb., 13 (1969), pp. 180-190.
5. E. Hopf, The general temporally discrete Markov process, J. Math. Mech., 3 (1954), pp. 13-45.
6. E. M. Klimko, On spreading partitions and entropy in infinite measure spaces, unpublished, partially contained in [7].
7. E. M. Klimko, On the entropy of a product endomorphism in infinite measure spaces, Z. Wahrscheinlichkeitstheorie verw. Geb., 12 (1969), pp. 290-292.
8. E. M. Klimko and L. Sucheston, On convergence of information in spaces with invariant measure, Z. Wahrscheinlichkeitstheorie verw. Geb., 10 (1968), pp. 226-235.
9. A. N. Kolmogorov, A new metric invariant of transient dynamical systems and automorphisms of Lebesgue spaces, Dokl. Akad. Nauk SSSR, 119 (1958), pp. 861-864. (Russian.)
10. U. Krengel, Entropy of conservative transformations, Z. Wahrscheinlichkeitstheorie verw. Geb., 7 (1967), pp. 161-181.
11. U. Krengel, Darstellungsätze für Stömungen und Halbströmungen, II, Math. Ann., 182 (1969), pp. 1-39.
12. U. Krengel, Transformations without finite invariant measure have finite strong generators, to appear.
13. W. Krieger, On entropy and generators of measure preserving transformations, Trans. Amer. Math. Soc., to appear.
14. J. Neveu, Une demonstration simplifièe et une extension de la formule d'Abramov sur l'entropie des transformations induites, Z. Wahrscheinlichkeitstheorie verw. Geb., 13 (1969), pp. 135-140.
15. W. Parry, Entropy and Generators in Ergodic Theory, Math. Lecture Note Series, Benjamin, New York, (1969).

16. V. A. Rokhlin, Generators in ergodic theory II, Vestnik. Leningrad Univ. Mat. Meh. Astronom. (1965), pp. 68-72. (Russian.)

17. H. Scheller, Induzierte dynamische Systeme, Diplomarbeit der Mathematisch-Naturwissenschaftlichen Fakultät der Univ. Gottingen, 62 pages (1965).

SPECTRAL PROPERTIES OF MEASURE PRESERVING TRANSFORMATIONS

R. V. Chacon[1] – *University of Minnesota*

§1. INTRODUCTION. Let $(X, \mathcal{E}, \mu)$ be a nonatomic Lebesgue measure space and let τ be an invertible measure preserving transformation. The transformation τ induces a unitary transformation $T(\tau)$ of $\mathcal{L}_2(X, \mathcal{E}, \mu)$ obtained by setting $T(\tau)f(x) = f(\tau x)$. By the spectral properties of τ we mean the spectral properties of $T(\tau)$ restricted to the orthogonal complement of the constant functions, which is clearly strictly invariant.

There are a few results from operator theory which we need. If U is a unitary operator on a separable Hilbert space H, then for each $f \in H$ there is a unique measure μ_f on the unit circle C of the complex plane such that

$$(1) \qquad (U^n f, U^m f) = \int_C z^{n-m} \mu_f(dz),$$

$n, m = 0, \pm 1, \ldots$. For each $f \in H$ there is a minimal strictly invariant subspace $H(f)$ containing f, the subspace which is obtained by putting $H(f) = \overline{\{\ldots, U^{-1}f, f, Uf, \ldots\}}$; and if $H(f) = H$, then f is said to be a ***generator*** of U. Operators which have generators are called ***cyclic***, and if f and g are generators, then $\mu_f \cong \mu_g$.

We have furthermore that if U is unitary, then U can be written as the orthogonal sum of cyclic unitary operators $U = U_1 \oplus U_2 \oplus \ldots$, with $H = H_1 \oplus H_2 \oplus \ldots$, U_i acting on H_i, and with the corresponding measures being restrictions of the preceding ones, i.e. there exist Borel subsets $\{E_n\}$, $C \supset E_n \supset E_{n+1}$, such that $\mu_i(A) = \mu_1(A \cap E_i)$, where $\mu_i = \mu_{f_i}$, and we have also that the sequence is unique up to equivalence.

The spectral multiplicity of the unitary operator U is defined in terms of the sequence $\{\mu_i\}$: we say that the spectrum has ***multiplicity k*** if $\mu_1 \neq 0, \ldots, \mu_k \neq 0$, $\mu_{k+1} = 0$. If the spectrum has multiplicity one, then U is said to have ***simple spectrum.***

There is a final result from operator theory which is needed; we state it as follows.

THEOREM 1.1. If U is a unitary operator on a (separable) Hilbert space H and if the spectrum of U is not simple, then there exist orthonormal u, v in H such that

[1] Research supported in part by the National Science Foundation (GP 1177-2).

$$d^2(u, H(w)) + d^2(v, H(w)) \geqslant 1$$

for any $w \in H$.

Proof. Since the spectrum of U is not simple, $U = U_1 \oplus U_2 \oplus \ldots$, where $\mu_1 \neq 0$ and $\mu_2 \neq 0$, and since μ_2 is a restriction of μ_1, there exists a Borel set E such that $\mu_1(E) = \mu_2(E) > 0$. We let $\tilde{u}$ and $\tilde{v}$ be the vectors in H_1 and H_2 corresponding to the characteristic function of E under the isomorphism implicit in (1), and let $u = \tilde{u}/\|\tilde{u}\|$, $v = \tilde{v}/\|\tilde{v}\|$.

We have that $d^2(u, H(w)) \geqslant \inf d^2(u, g) - \epsilon/2$, $d^2(v, H(w)) \geqslant \inf d^2(v, g) - \epsilon/2$, where g is of the form $g = P(T, T^{-1})[Q(T, T^{-1})f_1 + R(T, T^{-1})f_2]$ with P an arbitrary polynomial, Q and R fixed polynomials, and f_1 and f_2 the generators of H_1 and H_2. This implies that

$$d^2(u, H(w)) \geqslant \int |a\chi_E - P_1 Qa\chi_E|^2 + |P_1 Ra\chi_E|^2 - \epsilon,$$

$$d^2(v, H(w)) \geqslant \int |P_2 Qa\chi_E|^2 + |a\chi_E - P_2 Ra\chi_E|^2 - \epsilon,$$

for each $\epsilon > 0$, for polynomials P_1, P_2, Q and R, where $a^2 = 1/\mu(E)$. The theorem follows on noting that

$$|1 - P_1 Q|^2 + |P_1 R|^2 + |P_2 Q|^2 + |1 - P_2 R|^2 \geqslant 1,$$

as we see by first remarking that we may assume without loss of generality that P_1, P_2, Q and R are positive. The sum $(1 - P_1 Q)^2 + (P_1 R)^2$ is the square of the distance between $(1, 0)$ and $(P_1 Q, P_1 R)$, and the sum $(P_2 Q)^2 + (1 - P_2 R)^2$ is the square of the distance between $(0, 1)$ and $(P_2 Q, P_2 R)$. The sum of these squares dominates the sum of the squares of the distances from $(0, 1)$ and $(1, 0)$ to the line through the origin having slope R/Q, which is clearly greater than or equal to 1.

The results outlined here are valid, with certain modifications, for normal operators, and are well-known. Theorem 1.1 is a known result which is proved here for completeness, and which is stated (without proof and in weaker form) in [9].

§2. TRANSFORMATIONS WITH SIMPLE SPECTRUM. In this section we study two classes of invertible measure preserving transformations which have simple spectrum, and which were introduced in [3].

DEFINITION 2.1. A *partition* $\xi = \{C_0, \ldots, C_q\}$ is a collection of pairwise disjoint measurable sets. If $A \in \mathcal{E}$, by $A(\xi)$ we mean the set consisting of all sets of ξ with the property that $\mu(A \,\Delta\, A(\xi))$ is minimal subject to this restriction.

DEFINITION 2.2. A sequence $\{\xi(n)\}$ of partitions *converges to the unit partition*, and we write $\xi(n) \to \epsilon$, provided that for each measurable set A, $\mu(A \,\Delta\, A(\xi(n))) \to 0$.

In the ~~algebra~~ ring $\tilde{\xi}$ of sets in $\mathcal{E}$ generated by ξ, there will be sets M such that $\mu(A \Delta M) = \min\{\mu(A \Delta K) : K \in \tilde{\xi}\}$; $A(\xi)$ stands for any of these sets M.

[$\tilde{\xi}$ is the family consisting of all unions of (finite) subsets of ξ.]

DEFINITION 2.3. The *class* $\mathcal{A}$ consists of those invertible measure preserving transformations τ for which there exists a sequence $\{\xi(n)\}$ of partitions such that

i. $\xi(n) \to \epsilon$;

ii. $\tau C_{k-1}(n) = C_k(n)$, $k = 1, \ldots, q(n)$.

The *class* $\mathcal{B}$ consists of those invertible measure preserving transformations τ in $\mathcal{A}$ such that

iii. $\lim_{n\to\infty} \mu(\tau C_{q(n)}(n) \Delta C_1(n))/\mu(C_1(n)) = 0.$

We first discuss a theorem obtained by J. Baxter [2].

LEMMA 2.1. If $\tau \in \mathcal{A}$, then τ has simple spectrum.

Proof. Let $\{\xi(n)\}$ be a sequence of partitions satisfying (i) and (ii). If $\chi_{C_0(n)}$ is the characteristic function of $C_0(n)$, then the subspace $H(\chi_{C_0(n)})$ has the property that if P_n is its orthogonal projection, then $P_n \to I$ (recall that if $\xi(n) = \{C_0(n), \ldots, C_{q(n)}(n)\}$ then $\tau C_{k-1}(n) = C_k$, $k = 1, \ldots, q(n)$, and that $\xi(n) \to \epsilon$). It then follows by Theorem 1.1 that τ has simple spectrum.

THEOREM 2.1 [2]. If $\tau \in \mathcal{A}$, let $\{\xi(n)\}$ be a sequence of partitions satisfying (i) and (ii). Then there exists a function v of the form

$$v = \sum_{k=1}^{\infty} \chi_{C_0}(n_k)$$

such that $H(v) = \mathcal{L}_2$, where χ_C is the characteristic function of C. Furthermore, the norm of v can be chosen arbitrarily small and the sets $C_0(n_k)$ pairwise disjoint.

Proof. By Lemma 2.1, τ has simple spectrum and the projections P_n on $H(\chi_{C_0(n)})$ satisfy $P_n \to I$. It follows therefore that there exist corresponding functions on the unit circle of the complex plane whose support has measure tending to the spectral measure of the circle. This implies ($\|\chi_{C_0(n)}\| \to 0$) that there is a subsequence $\{n_k\}$ such that the function corresponding to

$$v = \sum_{k=1}^{\infty} \chi_{C_0(n_k)}$$

has support on a set supporting the spectral measure, and we have $H(v) = \mathcal{L}_2$. We can also clearly make the norm of v as small as desired and the sets $C_0(n_k)$ pairwise disjoint (by modifying the sequence $\{\xi(n)\}$).

The class of transformations $\mathcal{A}$ is therefore seen to have a particularly simple spectrum, and this is borne out by the following results.

THEOREM 2.2 [1]. If $\tau \in \mathcal{A}$ and $\sigma\tau = \tau\sigma$, then there exist two sequences $\{X_n^1\}, \{X_n^2\}$ of sets and two sequences of nonnegative integers $\{j_n^1\}, \{j_n^2\}$ such that $X_n^1 \cap X_n^2 = \emptyset$ and such that

(i) $A \in \mathcal{E}$ implies that

$$\sigma(A) = \lim_{n\to\infty} \{ \tau^{j_n^1}(A \cap X_n^1) + \tau^{j_n^2}(A \cap X_n^2)\}$$

and

(ii) either $\sigma = \tau^k$ for some k, or

$$\lim_{n\to\infty} j_n^1 = \lim_{n\to\infty} j_n^2 = \infty.$$

A stronger theorem may be obtained for the second class:

THEOREM 2.3 [6]. If $\tau \in \mathcal{B}$ and $\sigma\tau = \tau\sigma$, then there is a sequence $\{j_n\}$ of nonnegative integers such that $A \in \mathcal{E}$ implies that

$$\sigma(A) = \lim_{n\to\infty} \tau^{j_n}(A).$$

The proofs of these theorems will be given in several lemmas. We first state some additional definitions.

DEFINITION 2.4. Given an invertible measure preserving transformation τ of X we say that a partition ξ is *τ-admissible* if $\xi = \{A_0, A_1, \ldots, A_q\}$ is finite and if $\tau A_{i-1} = A_i$, $1 \leqslant i \leqslant q$.

In what follows, τ and σ are two commuting invertible measure preserving transformations of X. For $A, B \in \mathcal{E}$, we write $A = B + E(\epsilon)$ provided that $\mu(A \,\Delta\, B) \leqslant \epsilon$.

LEMMA 2.2. Let $\xi = \{A_0, A_1, \ldots, A_q\}$ be τ-admissible, $A, B \in \mathcal{E}(\xi)$, and let $\sigma A = B + E(\epsilon)$. Then there are two disjoint sets F and G, and two nonnegative integers j, k, such that $\sigma A = \tau^j(A \cap F) + \tau^{-k}(A \cap G) + E(2\epsilon)$, $j + k \geqslant q + 1$ and $\mu(F) \leqslant k/(q+1)$, $\mu(G) \leqslant j/(q+1)$.

Proof. Let $C_i = A_0 \cap \sigma\tau^i A_0$, $-q \leqslant i \leqslant q$. If $x \in C_i \cap C_j$, then there exist two points, $y, z \in A_0$ such that $x = \sigma\tau^i y = \sigma\tau^j z$, which implies that $\tau^{i-j} y = z$. Hence either $i = j$ or $|i - j| \geqslant q + 1$. Therefore any $q + 1$ consecutive C_i's are pairwise disjoint.

Now if we let $R = X - X_\xi$, then for any ℓ, $0 \leqslant \ell \leqslant q$,

$$\sigma A_\ell = \sum_{k=0}^{q} A_k \cap \sigma A_\ell + R \cap \sigma A_\ell = \sum_{k=0}^{q} \tau^k C_{\ell-k} + R \cap \sigma A_\ell.$$

Hence by writing

$$A = \sum_{\ell=0}^{q} a_\ell A_\ell, \quad B = \sum_{k=0}^{q} b_k A_k = \sum_{k=0}^{q} \tau^k b_k A_0,$$

we obtain

$$\sigma A = \sum_{\ell=0}^{q} \tau^k \sum_{\ell=0}^{q} a_\ell C_{\ell-k} + R \cap \sigma A = \sum_{k=0}^{q} \tau^k b_k A_0 + E(\epsilon),$$

or

$$\sum_{k=0}^{q} \tau^k \sum_{\ell=0}^{q} a_\ell C_{\ell-k} = \sum_{k=0}^{q} \tau^k b_k A_0 + E(\epsilon).$$

Since for a fixed k the sets $C_{\ell-k}$, $0 \leqslant \ell \leqslant q$, are pairwise disjoint, we have

$$\sum_{k=0}^{q} \int_{A_0} \Big| \sum_{\ell=0}^{q} a_\ell \chi_{\ell-k} - b_k \Big| d\mu < \epsilon,$$

where χ_i is the characteristic function of C_i.

Therefore there exists a point $x_0 \in A_0$ such that

$$\sum_{k=0}^{q} \Big| \sum_{\ell=0}^{q} a_\ell \chi_{\ell-k}(x_0) - b_k \Big| \mu(A_0) < \epsilon,$$

which means that $D = B + E(\epsilon)$, where

$$D = \sum_{k=0}^{q} \sum_{\ell=0}^{q} a_\ell \chi_{\ell-k}(x_0) A_k.$$

Hence we have that $\sigma A = D + E(2\epsilon)$. To see that this implies the lemma, we rearrange the double summation for D as follows:

$$D = \chi_0(x_0)A + \sum_{j=1}^{q} \chi_{-j}(x_0)\tau^j \sum_{\ell=0}^{q-j} a_\ell A_\ell + \sum_{k=1}^{q} \chi_k(x_0)\tau^{-k} \sum_{\ell=k}^{q} a_\ell A_\ell,$$

where there are at most two nonzero terms. If there are less than two nonzero terms, we may define $F = \emptyset$ or $G = \emptyset$ or $F = G = \emptyset$ and see that all the conclusions of the lemma are satisfied. If there are exactly two nonzero terms, then

$$D = \tau^j \sum_{\ell=0}^{q-j} a_\ell A_\ell + \tau^{-k} \sum_{\ell=k}^{q} a_\ell A_\ell,$$

where $j, k \geqslant 1$ and $j + k \geqslant q + 1$, since $C_{-j} \cap C_k \neq \emptyset$ implies $j + k \geqslant q + 1$ because any $q + 1$ consecutive C_i's are pairwise disjoint. In this case we let $F = \sum_{\ell=0}^{q+j} A_\ell$, $G = \sum_{\ell=k}^{q} A_\ell$, and we have

$$\mu(F) \leqslant [(q - j + 1)/(q + 1)]\,\mu(X_\xi) \leqslant k(q + 1);$$

similarly $\mu(G) \leqslant j/(q + 1)$.

We may now prove Theorem 2.2.

Proof of Theorem 2.2. Let $\xi_n = \{A_0(n), A_1(n), \ldots, A_{q(n)}(n)\}$, $\xi_n \to \epsilon$, be a sequence of τ-admissible partitions. Note that $\lim_{n\to\infty} q(n) = \infty$. Choose a sequence $\epsilon_n > > 0$ such that $\epsilon_n q(n) \to 0$ as $n \to \infty$. For any integer $n = 1, 2, \ldots$, we can find another integer $m(n)$ with the following properties:

(a) there exist $B_n, D_n \in \mathcal{E}(\xi_{m(n)})$ such that $A_0(n) = B_n + E(\epsilon_n)$, $\sigma A_0(n) = = D_n + E(\epsilon_n)$, and

(b) $q(n) \leqslant \epsilon_n q(m(n))$.

We then have that $\sigma B_n = D_n + E(2\epsilon_n)$. Now choose j_n^1 and j_n^2 and F_n, G_n according to Lemma 2.2, with respect to the partitition $\xi_{m(n)}$ and the set $B_n \in \mathcal{E}(\xi_{m(n)})$. Hence $\sigma B_n = \tau^{j_n^1}(B_n \cap F_n) + \tau^{-j_n^2}(B_n \cap G_n) + E(4\epsilon_n)$, and therefore

$$\sigma A_0(n) = \tau^{j_n^1}(A_0(n) \cap F_n) + \tau^{-j_n^2}(A_0(n) \cap G_n) + E(7\epsilon_n).$$

If we now let $X_n^1 = \sum_{k=0}^{q(n)} \tau^k(A_0(n) \cap F_n)$ and $X_n^2 = \sum_{k=0}^{q(n)} \tau^k(A_0(n) \cap G_n)$, we have that for any $A \in \mathcal{E}(\xi_n)$, $\sigma A = \tau^{j_n}(A \cap X_n^1) + \tau^{-j_n^2}(A \cap X_n^2) + E(7(q(n) + 1)\epsilon_n)$, and it is clear that this implies (i).

To prove (ii), we note for example that $\mu(X_n^2) \leqslant (q(n) + 1)\mu(G_n) \leqslant (q(n) + + 1/q(n(m)) + 1)j_n^1$, which means that $\liminf_{n\to\infty} \mu(X_n^2) \leqslant \liminf_{n\to\infty} (\epsilon_n j_n^1)$. Therefore if j_n^1 has a bounded subsequence, then by considering a subsequence if necessary, one may assume that $X_n^2 = \emptyset$ and $j_n^1 = k$ for all n.

As an application of this theorem, we give the following result.

THEOREM 2.4 [1]. If a mixing transformation $\tau \in \mathcal{Q}$, then it can commute only with its powers.

Before giving the proof, we note that the nontrivial fact that there are mixing transformations that have an approximation by partitions follows from the results given in [10]. Hence Theorem 2.4 is not vacuously true.

Proof. Using the notation of Theorem 2.2, we have $\sigma A \subset \tau^{j_n^1} A + \tau^{-j_n^2} A + E_n$ for all $n \geqslant 1$ where $\mu(E_n) \to 0$. Hence

$$\sigma A \subset \{\tau^{j_n^1} A + \tau^{-j_n^2} A + E_n\} \cap \{\tau^{j_m^1} A + \tau^{-j_m^2} A + E_m\},$$

or

$$\sigma A \subset \tau^{j_n^1} A \cap \tau^{j_m^1} A + \tau^{j_n^1} A \cap \tau^{-j_m^2} A + \tau^{-j_n^2} A \cap \tau^{j_m^1} A + \tau^{-j_n^2} A \cap \tau^{-j_m^2} A + E_{mn}$$

where $\mu(E_{mn}) \to 0$ as $m, n \to \infty$. Now, if σ is not a power of τ, then $j_n^1, j_n^2 \to \infty$ as $n \to \infty$. Hence, letting first $n \to \infty$ then $m \to \infty$, we see, from the mixing property,

$$\mu(\sigma A) = \mu(A) \leqslant 4[\mu(A)]^2 \quad \text{for all } A \in \mathcal{E},$$

which is a contradiction.

LEMMA 2.3. If, in addition to the hypotheses of Lemma 2.2, $A_0 = \tau A_q + E(\delta)$, then there exists an integer $j, -q \leqslant j \leqslant q$ and a set $A' \subset A$ such that $\sigma A = \tau^j A' + E(2\epsilon + 2(q+1)\delta)$.

Proof. We already have that $\sigma A = D + E(2\epsilon)$. If $D = \emptyset$ or D contains only one nonzero term, the lemma follows. Also, if $D = \tau^j \sum_{\ell=0}^{q-j} a_\ell A_\ell + \tau^{-k} \sum_{\ell=k}^{q} a_\ell A_\ell$, where $j + k = q + 1$, then we may write $\tau^{-k} A_\ell = \tau^j \tau^{-q-1} A_\ell = \tau^j A_\ell + E(\delta)$, which means that $D = \tau^j A + E((q+1)\delta)$, and again the lemma follows. Therefore, the only case which need be considered is the case for which $\chi_{-j}(x_0) + \chi_k(x_0) = 2$ for some j, $k \geqslant 1$ with $j + k > q + 1$.

We will now give an upper bound for the measure of the set

$$S = \{x: \exists\, j, k,\ 1 \leqslant j, k \leqslant q,\ 1 + q < j + k,\ \chi_{-j}(x) + \chi_k(x) = 2\}.$$

By the definition of S,

$$\begin{aligned} S &= \sum_{j=2}^{q} \sum_{k=q-j+2}^{q} C_{-j} \cap C_k = \sum_{j=2}^{q} \sum_{k=q-j+2}^{q} A_0 \cap \tau^{-j}\sigma A_0 \cap \tau^k \sigma A_0 \\ &= \sum_{j=2}^{q} \tau^{-j}\{A_j \cap \sigma(A_0 \cap \sum_{k=q-j+2}^{q} \tau^{k+j} A_0)\} \\ &= \sum_{j=2}^{q} \tau^{-j}\{A_j \cap \sigma(A_0 \cap \sum_{\ell=2}^{q} \tau^{\ell} A_q)\} \end{aligned}$$

which means that $\mu(S) \leqslant \mu(A_0 \cap \sum_{\ell=1}^{q-1} \tau^{1+\ell} A_q)$. Hence $\mu(A_0 \cap \tau A_q) > \mu(A_0) - \delta$ implies that $\mu(S) < \delta$.

Now go back to

$$\int_{A_0} \sum_{k=0}^{q} \Big| \sum_{\ell=0}^{q} a_\ell \chi_{\ell-k} - b_k \Big| d\mu < \epsilon;$$

since the integrand is bounded by $q + 1$, there exists a point $x_0 \in A_0 - S$ such that

$$\sum_{k=0}^{q} \Big| \sum_{\ell=0}^{q} a_\ell \chi_{\ell-k}(x_0) - b_k \Big| \mu(A_0) < \epsilon + (q+1)\delta.$$

Proof of Theorem 2.3. The proof is the same as the proof of Theorem 2.2 with Lemma 2.3 replacing Lemma 2.2.

§3. APPROXIMATION OF MEASURABLE TRANSFORMATIONS. Although we are interested in the class $\mathcal{C}$(MM) of invertible, measurable and measure preserving transformations, it turns out to be more convenient to consider a wider class of transformations, the nonsingular measurable transformations which are one-to-one from their domain to their range. We denote this class of transformations by $\mathcal{C}(M)$ and define the class more formally as follows.

DEFINITION 3.1. $\tau \in \mathcal{C}(M)$ if there exist sets $D(\tau)$ and $R(\tau)$ contained in X, the domain and range sets of τ, such that:

(i) $\tau: D(\tau) \rightarrow R(\tau)$ is one-to-one and onto;

(ii) if $A \in \mathcal{E}$, then $\tau(A)$ and $\tau^{-1}(A) \in \mathcal{E}$, and $D(\tau)$ (and $R(\tau)$) is in $\mathcal{E}$;

(iii) if $A \in \mathcal{E}$ and $A \subset D(\tau)$, then $\mu(\tau(A)) = 0$ if and only if $\mu(A) = 0$;

(iv) if $A \in \mathcal{E}$ and $A \subset R(\tau)$, then $\mu(\tau^{-1}(A)) = 0$ if and only if $\mu(A) = 0$.

DEFINITION 3.2. Let $\tau_1, \tau_2 \in \mathcal{C}(M)$ and let $\xi = \{C_0, \ldots, C_q\}$ be a partition. We define the distance between τ_1 and τ_2 with respect to the partition ξ as

$$d_\xi(\tau_1, \tau_2) = \sum_{k=0}^{q} \mu(\tau_1 C_k \,\Delta\, \tau_2 C_k).$$

By τ_ξ we mean any transformation in $\mathcal{C}(M)$ with

(i) $D(\tau_\xi) = R(\tau_\xi) = \bigcup_{k=0}^{q} C_k$;

(ii) $\tau_\xi C_{k-1} = C_k$, $k = 1, \ldots, q$, $\tau_\xi C_q = C_1$,

where $\xi = \{C_0, \ldots, C_q\}$. It follows easily from the definition that $d_\xi(\tau_\xi, \tau)$ is uniquely defined for $\tau \in \mathcal{C}(M)$, independently of the particular τ_ξ chosen. It is also clear that if the sets of ξ have the same measure, then τ_ξ can be chosen in $\mathcal{C}(MM)$.

LEMMA 3.1. If $\mu(\tau_1 A \,\Delta\, \tau_2 A) = 0$ for all $A \in \mathcal{E}$, then $\tau_1 = \tau_2$ (almost everywhere).

Proof. There exists a denumerable family of measurable sets $\{E_n\}$ so that each point $x \in X$ is equal to the intersection of some of these sets. There also exists a denumerable family of measurable sets $\{E_n^i\}$, $i = 1, 2$, such that $E_n^i \subset E_n$, $\mu(E_n - E_n^i) = 0$ and $\tau_1(E_n^1) = \tau_2(E_n^2)$. This implies that almost all $x \in X$ have the property that each is equal to the same intersection but with the E_n's replaced by the E_n^i's:

$$x = \cap\{E_n^i : n \in \Lambda(x)\}, \quad i = 1, 2,$$

($\Lambda(x)$ is independent of

and therefore $\tau_1 x = \tau_2 x = \cap\{\tau_1(E_n^1): n \in \Lambda(x)\} = \cap\{\tau_2(E_n^2): n \in \Lambda(x)\}$.

One may introduce borel sets P and Q in X such that $P \subseteq D(\tau_1) \cap D(\tau_2)$, $Q \subseteq R(\tau_1) \cap R(\tau_2)$, $\tau_1(P) = Q$, $\tau_2(P) = Q$, $\mu(D(\tau_i) \setminus P) = 0$, $\mu(R(\tau_i) \setminus Q) = 0$, $i = 1, 2$.

LEMMA 3.2. If $\xi(n) \to \epsilon$ and if $d_{\xi(n)}(\tau_1, \tau_2) \to 0$, then $\tau_1 = \tau_2$.

Proof.

$$\mu(\tau_1 A \triangle \tau_2 A) \leqslant \mu(\tau_1 A \triangle \tau_1 A(\xi(n))) + \mu(\tau_1 A(\xi(n)) \triangle \tau_2 A)$$

$$\leqslant \mu(\tau_1 A \triangle \tau_1 A(\xi(n))) + \mu(\tau_2 A \triangle \tau_2 A(\xi(n))) + \mu(\tau_1 A(\xi(n)) \triangle \tau_2 A(\xi(n)))$$

$$\leqslant 2\mu(A \triangle A(\xi(n)) + d_{\xi(n)}(\tau_1, \tau_2).$$

DEFINITION 3.3. Let $\xi(n) = \{C_0(n), \ldots, C_{q(n)}(n)\}$, $n = 1, 2, \ldots$, be a sequence of partitions, and let τ and $\tau(n)$, $n = 1, 2, \ldots$, be in $\mathcal{C}(M)$. We say that τ admits of *approximation with speed f(n)* with respect to $\{\xi(n)\}$ and $\{\tau(n)\}$ if

(i) $\xi(n) \to \epsilon$, and

(ii) $d_{\xi(n)}(\tau, \tau(n)) \leqslant f(q(n))$.

If $\tau(n) = \tau_{\xi(n)}$, and if $\mu(C_i(n)) = \mu(C_1(n))$, $i = 1, \ldots, q(n)$, then we say that τ admits of *cyclic approximation with speed f(n)*.

The notion of cyclic approximation is given by Katok and Stepin in [9], and it can be given in a simpler way. This may be stated in the form of the following lemma.

LEMMA 3.3. τ admits of cyclic approximation with speed $f(n)$ if and only if there is a sequence of partitions $\xi(n) = \{C_0(n), \ldots, C_{q(n)}(n)\}$, and $\tau_{\xi(n)}$ such that

(i) $\xi(n) \to \epsilon$;

(ii) $\mu(\tau_{\xi(n)} \neq \tau) \leqslant f(n)/2$.

§4. APPROXIMATION OF TRANSFORMATIONS WITH CONTINUOUS SPECTRUM. In this section we study the connection between approximability and the continuity of the spectrum. We refer to [5] for the proofs of the principal results, limiting the discussion to an outline of the results obtained.

DEFINITION 4.1. We denote by $\mathcal{U}$ the class of invertible measure preserving transformations whose domain and range equal X.

LEMMA 4.1. The set $\mathcal{U}$ is a topological group with respect to the weak topology, i.e. the topology obtained by taking neighborhoods to be finite intersections of sets of the form $\{\sigma\colon \sigma \in \mathcal{U}\ \ \mu(\sigma E \triangle \tau E) < \epsilon\}$, $E \in \mathcal{E}$.

In what follows it will be convenient to let $0 \cdot B = \emptyset$, $1 \cdot B = B$, for sets $B \in \mathcal{E}$.

In what follows also, the topological space we refer to is $\mathcal{U}$, equipped with the weak topology.

DEFINITION 4.2. We say that ξ is a *tower* if $\xi = \{C_i, i = 0, \ldots, q\}$ is a partition of sets of the same measure whose union $A = A_\xi \subset X$.

DEFINITION 4.3. Let $\xi = \{C_i, i = 0, \ldots, q\}$ be a tower, and let $\tau \in \mathcal{U}$. We say that σ maps the elements of the tower *almost cyclically* if there exists an integer m, $0 \leqslant m \leqslant q$ such that the collection of sets $\{\sigma^k C_m, k = 0, \ldots, q\}$ is the same as the collection (without order) of sets which make up the tower ξ.

DEFINITION 4.4. Let $\xi(n)$ be a sequence of towers. By $\mathcal{M}(\{\xi(n)\})$ we mean the class of $\sigma \in \mathcal{U}$ with the property that there exists an integer N(S) such that for $n \geqslant N(S)$, σ maps the elements of $\xi(n)$ almost cyclically.

DEFINITION 4.5. Let $\xi(1)$ and $\xi(2)$ be partitions, $\xi(1) = \{C_{1,i}, i = 0, \ldots, q(1)\}$ and $\xi(2) = \{C_{2,i}, i = 0, \ldots, q(2)\}$. We write $\xi(1) < \xi(2)$ if each set of $\xi(1)$ is the sum of the sets of $\xi(2)$. We say that a sequence of partitions $\{\xi(n)\}$ is *nested* if $\xi(n) < \xi(n+1)$, $n = 1, 2, 3, \ldots$.

LEMMA 4.2 [5]. Let $\{\xi(n)\}$ be a nested sequence of towers such that $\xi(n) \to \epsilon$ as $n \to \infty$. Then the set $\mathcal{M}(\{\xi(n)\})$ is dense in $\mathcal{U}$.

DEFINITION 4.6. A sequence $\{n(k)\}$ is called an *m-pair sequence* if

$$n(2k) = 1 + m \cdot n(2k-1), \quad k = 1, 2, 3, \ldots .$$

We say that $\tau \in \mathcal{U}$ *admits of approximation in m-pairs with speed f(n)* if it admits of cyclic approximation with speed f(n), and if the sequence $\{q(n)\}$, where q(n) is the number of elements in $\xi(n)$, has a subsequence which is an m-pair sequence.

We refer to the elements of the group $\mathcal{U}$ as *automorphisms* and state several results obtained by Katok and Stepin in [10].

THEOREM 4.1. The set of automorphisms admitting of approximation with a fixed speed f(n) contains a G_δ set which is everywhere dense in $\mathcal{U}$.

THEOREM 4.2. If an automorphism admits of approximation with speed θ/n, $\theta < 4$, then it is ergodic, and if $\theta < 1/2$, then its spectrum is simple.

THEOREM 4.3. If an automorphism admits of approximation with speed θ/n, $\theta < 2$, then it is not strongly mixing.

THEOREM 4.4. Strongly mixing automorphisms form a set of the first category in the group $\mathcal{U}$.

Results analogous to Theorems 4.1, 4.3, and 4.4 were obtained in [5] for automorphisms with continuous spectrum, or what is the same thing, for automorphisms which are weakly mixing. Theorem 4.6 is a key result, as it implies that the theory is not vacuous and is used in the proofs of the remaining results. We give a proof of Theorem 4.5 to illustrate some of the ideas used in the proofs.

THEOREM 4.5 [5]. If the automorphism τ admits of approximation in m-pairs with speed θ/n, $\theta < 2$, then τ is weakly mixing.

Proof. For any integer k we have that $\mu(\tau^k \neq \tau_n^k), \leqslant k\mu(\tau \neq \tau_n)$, so that $\mu(\tau^{q(n)} \neq \tau_n^{q(n)}) \leqslant \theta$ (to see this for $k = 2$, note that

$$\mu(\tau^2 \neq \tau_n^2) = \mu(\tau = \tau_n,\ \tau^2 \neq \tau_n^2) + \mu(\tau \neq \tau_n,\ \tau^2 \neq \tau_n^2)$$

$$= \mu(\tau = \tau_n,\ \tau\tau_n \neq \tau_n\tau_n) + \mu(\tau \neq \tau_n,\ \tau^2 \neq \tau_n^2)$$

$$\leqslant \mu(\tau\tau_n \neq \tau_n\tau_n) + \mu(\tau \neq \tau_n) = \mu(\tau \neq \tau_n) + \mu(\tau \neq \tau_n)).$$

This implies that

$$\mu(\tau^{q(n)} = \tau_n^{q(n)}) \geqslant (1 - \theta)(1 - \delta(n)) \tag{1}$$

since $\mu(\tau^{q(n)} = \tau_n^{q(n)}) = 1 - \mu(\tau^{q(n)} = \tau_n^{q(n)}) \geqslant 1 - \theta$, and we obtain equation (1) with $\delta(n) = \mu(D(n))$, and $D(n) = cA(n)$, where $A(n)$ is the union of the sets of $\xi(n)$. The sets of the tower $\xi(n)$, $C_{n,i}$, $i = 0, \ldots, q(n)$, and the complement of their union $D(n)$ form a complete partition of X, and we therefore have for each function $f(x)$ that

$$f(x) = \sum_{j=0}^{q(n)} f(x)\chi_{C_{n,j}}(x) + f(x)\chi_{D(n)}(x)$$

where χ_C is the characteristic function of C. Further, since $\xi(n) \to \epsilon$, given $\epsilon > 0$ there exists a set $E = E(n)$ contained in $A(n)$ and of measure less than ϵ, $q + 1$ constants $f_0, \ldots, f_{q(n)}$, and a function $\epsilon(x)$ such that

$$f(x) = \sum_{j=0}^{q(n)} (f_j + \epsilon(x))\chi_{C_{n,j}}(x) + f(x)\chi_{D(n)}(x) \tag{2}$$

where $|\epsilon(x)| < \epsilon$, $x \in cE$.

If we let $f(x)$ be an eigenfunction with eigenvalue λ we have that

$$\lambda^{q(n)} f(x) = \sum_{j=0}^{q(n)} (f_j + \epsilon'(x))\chi_{C'_{n,j}}(x) + f'(x)\chi_{D'(n)}(x), \tag{3}$$

where $\epsilon'(x) = \epsilon(\tau^{-q(n)}x)$, $C'_{n,j} = \tau^{q(n)}C_{n,j}$, $f'(x) = f(\tau^{-q(n)}x)$, and $D'(n) = \tau^{q(n)}D(n)$.

If we let $B(n) = \sum_{j=0}^{q(n)} \tau^{q(n)} C_{n,j} \cap C_{n,j}$, then equation (1) implies that $\mu(B(n)) \geqslant (1 - \theta)(1 - \delta(n))$, and it follows from equation (2) that for $x \in cE \cap B(n)$ we have

$$f(x) = \sum_{j=0}^{q(n)} (f_j + \epsilon(x))\chi_{C_{n,j} \cap C'_{n,j}}(x). \tag{4}$$

Further, equation (3) implies that for $x \in c\tau^{q(n)}E \cap B(n)$ we have

$$\lambda^{q(n)} f(x) = \sum_{j=0}^{q(n)} (f_j + \epsilon'(x))\chi_{C_{n,j} \cap C'_{n,j}}(x). \tag{5}$$

Equations (4) and (5) in turn imply that

$$\lambda^{q(n)} = \sum_{j=0}^{q(n)} [(f_j + \epsilon(x))/(f_j + \epsilon'(x))]\chi_{C_{n,j} \cap C'_{n,j}}(x), \tag{6}$$

on $F(n) = B(n) \cap cE \cap c\tau^{q(n)}E$. By choosing $\epsilon > 0$ sufficiently small we can make the measure of $F(n)$ as close to $1 - \theta$ as we like, and we also have that $|\epsilon(X)| \leqslant \epsilon$, $|\epsilon'(x)| \leqslant \epsilon$. Since $|f(x)| = 1$, we may assume that $|f_j| = 1$ (τ is ergodic, and we have assumed that $f(x)$ is an eigenfunction, so that $|f(x)| = 1$, as well as $|\lambda| = 1$) and so (6) implies that on $F(n)$

$$|\lambda^{q(n)} - 1| \leqslant 2\epsilon/(1 - \epsilon), \tag{7}$$

where $\mu(F(n)) > 0$.

Since we have assumed that $\{q(n)\}$ contains an m-pair sequence, we may apply (7) with $q(n) = k$ and with $q(n) = 1 + mk$ to obtain in the first case that ($|\lambda| = 1$)

$$\begin{aligned} |\lambda^{1+mk} - 1| &= |\lambda\lambda^{mk} - \lambda^{mk} + \lambda^{mk} - 1| \\ &= |\lambda^{mk}(\lambda - 1) + (\lambda^{mk} - \lambda^{(m-1)k}) + \ldots + (\lambda - 1)| \\ &\geqslant |\lambda - 1| - (2m\epsilon/(1 - \epsilon)), \end{aligned} \tag{8}$$

and in the second that

$$|\lambda^{1+mk} - 1| \leqslant 2\epsilon/(1 - \epsilon). \tag{9}$$

These two inequalities imply that

$$|\lambda - 1| \leqslant 2(m + 1)\epsilon/(1 - \epsilon),$$

from which it follows that $\lambda = 1$.

THEOREM 4.6 [5]. Let m be a positive integer and let $\theta > 1/m$. Then there exists an automorphism $\tau(m)$ admitting of approximation in m-pairs with speed $2\theta/n$.

THEOREM 4.7 [5]. The set of automorphisms admitting of approximation in m-pairs with speed $2\theta/n$, $\theta > 1/m$, contains an everywhere dense G_δ set.

THEOREM 4.8 [5]. The automorphisms admitting of approximation in m-pairs with $2\theta/n$, $\theta = 1/m$, are weakly mixing but not strongly mixing, and they contain an everywhere dense G_δ set.

REFERENCES

1. M. A. Akcoglu, R. V. Chacon and T. Schwartzbauer, Commuting transformations and mixing, Proc. Amer. Math. Soc., to appear.
2. J. Baxter, A class of ergodic automrophisms, Thesis, University of Toronto, 1969.
3. R. V. Chacon, A geometric construction of measure preserving transformations, Fifth Berkeley Symposium on Probability and Statistics (2), 2 (1967), pp. 335-360.
4. R. V. Chacon, Weakly mixing transformations which are not strongly mixing, to appear.
5. R. V. Chacon, Approximation of transformations with continuous spectrum, to appear.
6. R. V. Chacon and T. Schwartzbauer, Commuting point transformations, to appear.
7. P. R. Halmos, Lectures on ergodic theory, Math. Soc. of Japan, (1956), Chelsea.
8. A. B. Katok and A. M. Stepin, Approximation of ergodic dynamical systems by periodic transformations (Russian), Dokl. Akad. Nauk SSSR, 171 (1966), pp. 1268-1271, (Soviet Math. Dokl., 7 (1966), pp. 1638-1641).
9. A. B. Kaotk and A. M. Stepin, Approximations in ergodic theory (Russian), Uspekhi Math. Nauk (5), 22 (1967), pp. 81-106, (Russian Math. Surveys (5) 22 (1967), pp. 77-102).
10. D. S. Ornstein, A mixing transformation that commutes only with its powers, to appear.
11. V. A. Rokhlin, On the fundamental ideas of measure theory (Russian), Math. Sbor. 67 (1949), pp. 107-150, (Amer. Math. Soc. Transl. 71 (1952)).
12. V. A. Rokhlin, Lectures on the entropy theory of measure preserving transformations (Russian), Uspekhi Math. Nauk (5), 22 (1967), pp. 3-56, (Russian Math. Surveys (5), 22 (1967), pp. 1-52).

4. Proc. AMS 22 (1969) 559-562
The example is not essentially different from what is described in Friedman's lectures (Van Nostrand).

OPTIMAL CONDITIONING OF OPERATORS ON HILBERT SPACE

C. McCarthy[1] – *University of Minnesota*

ABSTRACT. Let $\mathcal{D}$ be a von Neumann algebra of operators on a Hilbert space $\mathcal{H}$ and denote by $\mathcal{U}$ the unitary group of $\mathcal{D}'$. For any two operators R, S on $\mathcal{H}$, define $c_{\mathcal{D}}(R, S) = \inf\{\|RD^{-1}\|\,\|DS\|: D \in \mathcal{D}\}$, $b_{\mathcal{U}}(R, S) = \sup\{\|RUS\|: U \in \mathcal{U}\}$. We prove that if $\mathcal{D}$ is abelian, $c_{\mathcal{D}}(R, S) \leqslant 2b_{\mathcal{U}}(R, S)$, and if $\mathcal{D}'$ is abelian or if $\mathcal{H}$ is finite dimensional, $c_{\mathcal{D}}(S^{-1}, S) \leqslant [b_{\mathcal{U}}(S^{-1}, S)]^2$. Examples show that some generalizations need not be valid.

We also show that if R and S are invertible, then the infimum defining $c_{\mathcal{D}}(R, S)$ is attained. This requires an analogue for noncommuting positive operators of the inequality between the arithmetic and the harmonic mean, which itself seems to be new.

§1. INTRODUCTION. In another paper [1], G. Strang and the present author studied the conditioning of matrices in the norm given by interpreting matrices in the usual way as operators on a finite-dimensional Hilbert space. This paper treats a number of questions which arise naturally from the considerations of [1]. To a large extent, the results are parallel and the proofs often similar to the finite-dimensional case; the principal difficulties are topological and arise from the infinite dimensionality of the Hilbert space. Nevertheless, we hope that the present general treatment may lead to a better understanding of conditioning. In order to make this paper substantially self-contained, there is unavoidably some overlap of material with [1].

We should like here to express our deep gratitude to G. Strang, whose continuing interest and encouragement provided the motivation for this work.

§2. NOTATION AND PRELIMINARIES. We begin with a complex Hilbert space $\mathcal{H}$ with inner product $(\cdot\,,\cdot)$ and norm $|\cdot|$; the norm of an operator on $\mathcal{H}$ is denoted by $\|\cdot\|$. We fix a von Neumann algebra $\mathcal{D}$ of operators on $\mathcal{H}$, and denote by $\mathcal{D}_0$ the set of invertible operators in $\mathcal{D}$; the elements of $\mathcal{D}_0$ will be called *conditioning operators.* $\mathcal{U}$ will denote a subgroup of the group of unitary operators in $\mathcal{D}'$, the commutant of $\mathcal{D}$; $\mathcal{U}$ is always to be sufficiently large so that DU = UD for all $U \in \mathcal{U}$ implies $D \in \mathcal{D}$, and most often will be all unitary operators in $\mathcal{D}'$.

[1] Supported by the National Science Foundation (GP 1236-1). In addition, it is a pleasure to thank the University of Sussex and the Battelle Institute, Geneva, for their hospitality.

Given a pair of operators R, S on $\mathcal{H}$, we define the quantities:

$$c_{\mathcal{D}}(R, S) = \inf \{\|RD^{-1}\| \, \|DS\|: D \in \mathcal{D}_0\}$$

the *optimal condition number* of R, S with respect to $\mathcal{D}$;

$$b_{\mathcal{U}}(R, S) = \sup \{\|RUS\|: U \in \mathcal{U}\}$$

the *bound* of R, S with respect to $\mathcal{U}$;

$$\rho_{\mathcal{D},\mathcal{U}}(R, S) = c_{\mathcal{D}}(R, S)/b_{\mathcal{U}}(R, S); \; \rho_{\mathcal{D},\mathcal{U}} = \sup_{R,S} \rho_{\mathcal{D},\mathcal{U}}(R, S).$$

If $c_{\mathcal{D}}(R, S) = \|R\| \, \|S\|$, we say that the pair R, S is *optimally conditioned* with respect to $\mathcal{D}$; an operator $D \in \mathcal{D}_0$ for which $c_{\mathcal{D}}(R, S) = \|RD^{-1}\| \, \|DS\|$ is called an *optimal conditioner* for R, S.

We collect a few trivial remarks as propositions.

2.1. PROPOSITION. Let λ, μ be nonzero scalars. Then $c_{\mathcal{D}}(\lambda R, \mu S) = |\lambda| \, |\mu| \, c_{\mathcal{D}}(R, S)$, $b_{\mathcal{U}}(\lambda R, \mu S) = |\lambda| \, |\mu| \, b_{\mathcal{U}}(R, S)$, and $\rho_{\mathcal{D},\mathcal{U}}(\lambda R, \mu S) = \rho_{\mathcal{D},\mathcal{U}}(R, S)$.

2.2. PROPOSITION. Let $D \in \mathcal{D}_0$, $U, V \in \mathcal{U}$. Then $c_{\mathcal{D}}(RD^{-1}U, VDS) = c_{\mathcal{D}}(R, S)$ and $b_{\mathcal{U}}(RD^{-1}U, VDS) = b_{\mathcal{U}}(R, S)$.

2.3. PROPOSITION. $\rho_{\mathcal{D},\mathcal{U}}(R, S) \geqslant 1$.

Proof. If $D \in \mathcal{D}_0$, $U \in \mathcal{U}$, then $U = D^{-1}UD$, and we have

$$\|RUS\| = \|RD^{-1}UDS\| \leqslant \|RD^{-1}\| \, \|U\| \, \|DS\| = \|RD^{-1}\| \, \|DS\|.$$

2.4. PROPOSITION. Let V, W be partial isometries with the range of R contained in the initial space of V and the range of S^* contained in the initial space of W. Then $c_{\mathcal{D}}(VR, SW) = c_{\mathcal{D}}(R, S)$ and $b_{\mathcal{U}}(VR, SW) = b_{\mathcal{U}}(R, S)$.

Proof. The condition on V yields $|VRx| = |Rx|$ for any $x \in \mathcal{H}$, so $\|VRD^{-1}\| = \|RD^{-1}\|$ and $\|VRUSW\| = \|RUSW\|$. The condition on V yields similarly $\|DSW\| = \|W^*S^*D^*\| = \|S^*D^*\| = \|DS\|$ and $\|RUSW\| = \|W^*S^*U^*R^*\| = \|S^*U^*R^*\| = \|RUS\|$.

2.5. PROPOSITION. $\rho_{\mathcal{D},\mathcal{U}} = \sup \{\rho_{\mathcal{D},\mathcal{U}}(R, S): R \geqslant 0, \; S \geqslant 0\}$.

Proof. Using the polar decomposition of R, S we may find partial isometries V, W which satisfy the hypotheses of Proposition 2.4, and such that $VR = (R^*R)^{1/2} \geqslant 0$ and $SW = (SS^*)^{1/2} \geqslant 0$. An application of that proposition yields $\rho_{\mathcal{D},\mathcal{U}}(R, S) = \rho_{\mathcal{D},\mathcal{U}}((R^*R)^{1/2}, (SS^*)^{1/2})$.

2.6. PROPOSITION. $c_{\mathcal{D}}(R, S) = \inf \{\|RD^{-1}\| \, \|DS\|: D \in \mathcal{D}_0, \; D > 0\}$.

Proof. Let $D = V(D^*D)^{1/2}$ be the polar decomposition of $D \in \mathcal{D}_0$. Since D is invertible, so is $(D^*D)^{1/2}$, and V is unitary; since $\mathcal{D}$ is von Neumann, $(D^*D)^{1/2} \in \mathcal{D}_0$. We have $\|RD^{-1}\| = \|R(D^*D)^{-1/2}V^*\| = \|R(D^*D)^{-1/2}\|$ and $\|DS\| = \|V(D^*D)^{1/2}S\| = \|(D^*D)^{1/2}S\|$.

§3. THE EXISTENCE OF AN OPTIMAL CONDITIONER FOR INVERTIBLE R, S. In this section we show that if R, S are both invertible, then there exists a $D_0 \in \mathcal{D}_0$ such that $c_{\mathcal{D}}(R, S) = \|RD_0^{-1}\| \, \|D_0 S\|$. In view of Proposition 2.2 (with $U = V = I$), the pair RD_0^{-1}, $D_0 S$ will then be optimally conditioned with respect to $\mathcal{D}$. From our proof it may also be seen that the set of squares of positive optimal conditioners for R, S: $\{D^2 \in \mathcal{D}_0: D > 0, \; c_{\mathcal{D}}(R, S) = \|RD^{-1}\| \, \|DS\|\}$ is convex, and that the set of normalized squares of positive optimal conditioners: $\{D^2 \in \mathcal{D}_0: D > 0, \; c_{\mathcal{D}}(R, S) = \|RD^{-1}\|^2 = \|DS\|^2\}$ is bounded, convex, and compact in the weak and strong operator topologies.

3.1. LEMMA. Let R, S be invertible. Then $c_{\mathcal{D}}(R, S) = \inf \{\|RD^{-1}\| \, \|DS\|: D \in \mathcal{D}_0, \|RD^{-1}\| = \|DS\|, m \leqslant D \leqslant M\}$ where $m = \|R^{-1}\|^{-3/2} \|S^{-1}\|^{-3/2} \|R\|^{-1} \|S\|^{-1}$, $M = \|R\|^{3/2} \|S\|^{3/2} \|R^{-1}\| \, \|S^{-1}\|$.

Proof. If $\|D^{-1}\| \, \|D\| > \|R^{-1}\| \, \|R\| \, \|S\| \, \|S^{-1}\|$, we have

$$\|R^{-1}\| \, \|RD^{-1}\| \, \|DS\| \, \|S^{-1}\| \geqslant \|D^{-1}\| \, \|D\| > \|R^{-1}\| \, \|R\| \, \|S\| \, \|S^{-1}\|$$

from which $\|RD^{-1}\| \, \|DS\| > \|R\| \, \|S\| \geqslant c_{\mathcal{D}}(R, S)$. Thus to determine the infimum which defines $c_{\mathcal{D}}(R, S)$, we need only consider those $D \in \mathcal{D}_0$ for which $\|D^{-1}\| \, \|D\| \leqslant \|R^{-1}\| \, \|R\| \, \|S\| \, \|S^{-1}\|$; by Proposition 2.6 we need consider only such D which are also positive. Since $\|R(\lambda D)^{-1}\| \, \|(\lambda D)S\| = \|RD^{-1}\| \, \|DS\|$ for any $\lambda = 0$, we need consider only λD where for each D, $\lambda > 0$ is chosen so that $\|R(\lambda D)^{-1}\| = \|(\lambda D)S\|$: $\lambda = \|RD^{-1}\|^{1/2} \|DS\|^{-1/2}$. With this choice of λ we have

$$\|\lambda D\| = \|RD^{-1}\|^{1/2} \|DS\|^{-1/2} \|D\| \leqslant \|R\|^{1/2} \|D^{-1}\|^{1/2} (\|S\| / \|D^{-1}\|)^{-1/2} \|D\|$$

$$\leqslant \|R\|^{1/2} \|S\|^{1/2} \|D^{-1}\| \, \|D\| \leqslant \|R\|^{3/2} \|S\|^{3/2} \|R^{-1}\| \, \|S^{-1}\| = M,$$

and similarly $\|(\lambda D)^{-1}\| \leqslant m^{-1}$.

Now the set of $D \in \mathcal{D}_0$ for which $m \leqslant D \leqslant M$ is compact in the weak operator topology, and the obvious way to find an optimal conditioner D_0 is to choose $D_n \in \mathcal{D}_0$, $m \leqslant D_n \leqslant M$, such that $\|RD_n^{-1}\| \, \|D_n S\|$ decreases to $c_{\mathcal{D}}(R, S)$, and then let D_0 be some sort of limit of the D_n. This procedure fails unless special care is taken, because operator multiplication need not be jointly continuous in the weak operator topology and there is no assurance that D_n will converge, let alone to D_0^{-1}. For this reason we have adopted a procedure which requires the analogue for positive operators of the inequality between the harmonic mean and the arithmetic mean. As this inequality does not seem to be in the literature, we digress to prove it here.

3.2. LEMMA. Let $A_1, \ldots, A_n$ be positive operators, $a_1, \ldots, a_n$ positive real numbers with $\sum_{\nu=1}^{n} a_\nu = 1$. Then, in the sense of positive definiteness,

$$\left(\sum_{\nu=1}^{n} a_\nu A_\nu^{-1}\right)^{-1} \leqslant \sum_{\nu=1}^{n} a_\nu A_\nu.$$

Proof. (a) Suppose $n = 2$, $A_1 = I$, $A_2 = A$. Let $E(\cdot)$ be the spectral resolution of A, and let x be any element of $\mathcal{H}$. Then

$$((a_1 I + a_2 A^{-1})^{-1} x, x) = \int_0^\infty (a_1 + a_2\lambda^{-1})^{-1}(E(d\lambda)x, x)$$

$$\leqslant \int_0^\infty (a_1 + a_2\lambda)(E(d\lambda)x, x) = ((a_1 I + a_2 A)x, x),$$

the inequality being a consequence of $(E(d\lambda)x, x) \geqslant 0$ and $(a_1 + a_2\lambda^{-1})^{-1} \leqslant a_1 + a_2\lambda$ for every $\lambda > 0$.

(b) In general for $n = 2$, apply (a) with $A = A_1^{-1/2} A_2 A_1^{-1/2}$ and replace x by $A_1^{1/2} x$ to obtain

$$((a_1 I + a_2 (A_1^{-1/2} A_2 A_1^{-1/2})^{-1})^{-1} A_1^{1/2} x, A_1^{1/2} x) \leqslant$$

$$\leqslant ((a_1 I + a_2 (A_1^{-1/2} A_2 A_1^{-1/2})) A_1^{1/2} x, A_1^{1/2} x),$$

which is the same as $((a_1 A_1^{-1} + a_2 A_2^{-1})^{-1} x, x) \leqslant ((a_1 A_1 + a_2 A_2)x, x)$.

(c) Proceed by induction. Suppose the lemma is true for n. Set $b_1 = a_1 + \ldots + a_n$, $b_2 = a_{n+1}$, $B_1 = ((a_1/b_1)A_1 + \ldots + (a_n/b_1)A_n)$, $B_2 = A_{n+1}$ and obtain from (b)

$$(b_1 B_1^{-1} + b_2 B_2^{-1})^{-1} \leqslant b_1 B_1 + b_2 B_2 = \sum_{\nu=1}^{n+1} a_\nu A_\nu.$$

The induction hypothesis is

$$\left(\sum_{\nu=1}^{n} (a_\nu/b_1) A_\nu^{-1}\right)^{-1} \leqslant \sum_{\nu=1}^{n} (a_\nu/b_\nu) A_\nu = B_1,$$

from which

$$\left(\sum_{\nu=1}^{n} (a_\nu/b_1) A_\nu^{-1}\right) \geqslant B_1^{-1}, \quad \left(\sum_{\nu=1}^{n} a_\nu A_\nu^{-1}\right) + b_2 B_2^{-1} \geqslant b_1 B_1^{-1} + b_2 B_2^{-1},$$

and thus

$$\left(\sum_{\nu=1}^{n} a_\nu A_\nu^{-1} + a_{n+1} A_{n+1}^{-1}\right)^{-1} \leqslant (b_1 B_1 + b_2 B_2^{-1})^{-1} \leqslant \sum_{\nu=1}^{n} a_\nu A_\nu.$$

We return now to the problem of showing that the infimum defining $c_p(R, S)$ is attained if R and S are both invertible. Let m, M be the constants of Lemma 3.1, and choose $D_n \in \mathcal{D}_0$ such that $m \leqslant D_n \leqslant M$ and $\|RD_n^{-1}\| = \|D_n S\| \leqslant (c_{\mathcal{D}}(R, S) + 1/n)^{1/2}$. Let K_n be the closed (in the weak operator topology) convex hull of the set of operators $\{D_\nu^2 : \nu \geqslant n\}$. For any convex combination $A = \sum_{\nu \geqslant n} a_\nu D_\nu^2$ we have

$$(1) \qquad \|S^*AS\| \leqslant \sum_{\nu \geqslant n} a_\nu \|S^* D_\nu S\| = \sum_{\nu \geqslant n} a_\nu \|D_\nu S\|^2 \leqslant c_{\mathcal{D}}(R, S) + 1/n,$$

and by Lemma 3.2,

$$(2) \quad \|RA^{-1}R^*\| \leqslant \|R(\sum_{\nu \geqslant n} a_\nu D_\nu^{-2})R^*\| \leqslant \sum_{\nu \geqslant n} a_\nu \|RD_\nu^{-1}\|^2 \leqslant c_{\mathcal{D}}(R, S) + 1/n.$$

Inequalities (1) and (2) persist for limits of such A in the strong operator topology, since norms are lower semi-continuous and inverses continuous in this topology, and hence throughout K_n since the weak and strong operator closures of convex sets agree. As each K_n is compact and $K_1 \supset K_2 \supset \ldots$, there exists an operator $\Delta \in \bigcap_{n=1}^{\infty} K_n$. We have $m^2 \leqslant \Delta \leqslant M^2$, $\|S^*\Delta S\| \leqslant c_{\mathcal{D}}(R, S) + 1/n$ and $\|R\Delta^{-1}R^*\| \leqslant c_{\mathcal{D}}(R, S) + 1/n$ for every n. Setting $D_0 = \Delta^{1/2}$, we have $\|RD_0^{-1}\| \, \|D_0 S\| \leqslant c_{\mathcal{D}}(R, S)$ as desired. We have therefore proved

3.3. THEOREM. Let R, S be invertible. Then there exists $D_0 \in \mathcal{D}_0$ such that $c_{\mathcal{D}}(R, S) = \|RD_0^{-1}\| \, \|D_0 S\|$.

The last theorem of this section shows that a bound for $\rho_{\mathcal{D},\mathcal{U}}(R, S)$ which is uniform over all invertible R, S also is a bound for $\rho_{\mathcal{D},\mathcal{U}}$.

3.4. LEMMA. Let R, S be given and set $R_\epsilon = (R^*R)^{1/2} + \epsilon I$, $S_\epsilon = (SS^*)^{1/2} + \epsilon I$ for $\epsilon \geqslant 0$. Then $\rho_{\mathcal{D},\mathcal{U}}(R, S) \leqslant \liminf_{\epsilon \to 0} \rho_{\mathcal{D},\mathcal{U}}(R_\epsilon, S_\epsilon)$.

Proof. Since $R_\epsilon^2 \geqslant R_0^2$, $S_\epsilon^2 \geqslant S_0^2$ for every $\epsilon > 0$, we have for every positive $D \in \mathcal{D}_0$ $D^{-1}R_\epsilon^2 D^{-1} \geqslant D^{-1}R_0^2 D^{-1}$ and $DS_\epsilon^2 D \geqslant DS_0^2 D$. Hence $\|R_\epsilon D^{-1}\| \geqslant \|R_0 D^{-1}\|$, $\|DS\| \geqslant \|DS_0\|$, and so $c_{\mathcal{D}}(R_\epsilon, S_\epsilon) \geqslant c_{\mathcal{D}}(R_0, S_0)$ for every $\epsilon > 0$.

Since $R \to R_0$ and $S \to S_0$ in the uniform operator topology, we have $b_{\mathcal{U}}(R_0, S_0) = \lim_{\epsilon \to 0} b_{\mathcal{U}}(R_\epsilon, S_\epsilon)$. It follows that $\rho_{\mathcal{D},\mathcal{U}}(R_0, S_0) \leqslant \liminf_{\epsilon \to 0} \rho_{\mathcal{D},\mathcal{U}}(R_\epsilon, S_\epsilon)$. By Proposition 2.4, $\rho_{\mathcal{D},\mathcal{U}}(R, S) = \rho_{\mathcal{D},\mathcal{U}}(R_0, S_0)$.

3.5. THEOREM. $\rho_{\mathcal{D},\mathcal{U}} = \sup\{\rho_{\mathcal{D},\mathcal{U}}(R, S)$: R, S positive, optimally conditioned, and of norm $1\}$.

Proof. Lemma 3.4 shows that $\rho_{\mathcal{D},\mathcal{U}}$ is determined by $\rho_{\mathcal{D},\mathcal{U}}(R, S)$ for positive invertible R, S. For invertible R, S an optimal conditioner D_0 exists by Theorem 3.3. By Proposition 2.2, $\rho_{\mathcal{D},\mathcal{U}}(RD_0^{-1}, D_0 S) = \rho_{\mathcal{D},\mathcal{U}}(R, S)$, so we need only consider those invertible pairs R, S which are optimally conditioned. For these $(R^*R)^{1/2}$ and $(SS^*)^{1/2}$ are also optimally conditioned and have the same $\rho_{\mathcal{D},\mathcal{U}}$ by Proposition 2.4. Proposition 2.1 shows that there is no loss of generality in taking R and S both of norm 1.

§4. FURTHER PROPERTIES OF OPTIMAL CONDITIONING. Throughout this section we assume that R and S are positive semi-definite and of norm 1; thanks to Theorem 3.5, this entails no essential loss of generality. We write the spectral resolutions of R, S as $R = \int_0^1 \lambda E(d\lambda)$, $S = \int_0^1 \lambda F(d\lambda)$, and we further set $E_a = \int_{a+}^1 E(d\lambda)$, $F_b = \int_{b+}^1 F(d\lambda)$. We will show that R, S are optimally conditioned if and only if the pair E_a, F_b is optimally conditioned for every $a < 1$, $b < 1$.

4.1. LEMMA. Let R, S be optimally conditioned and let $p \geq 2$ be an integral power of 2. Then R^p, S^p are optimally conditioned.

Proof. By induction, it is enough to prove the lemma for $p = 2$. Let $D \in \mathcal{D}_0$ be positive. Since R, S are optimally conditioned, $\|RD^{-1}\| \, \|DS\| \geq \|R\| \, \|S\|$. We have $\|RD^{-1}\|^2 = \|D^{-1}R^2D^{-1}\|$ = spectral radius of $D^{-1}R^2D^{-1}$, since $D^{-1}R^2D^{-1}$ is self-adjoint. Now $R^2D^{-2} = D(D^{-1}R^2D^{-1})D^{-1}$ is similar to $D^{-1}R^2D^{-1}$, thus has the same spectral radius, and hence we have $\|R^2D^{-2}\| \geq \|RD^{-1}\|^2$. Similarly we have $\|D^2S^2\| \geq \|DS\|^2$ and so $\|R^2D^{-2}\| \, \|D^2S^2\| \geq \|R\|^2 \, \|S\|^2 = \|R^2\| \, \|S^2\|$ for every positive $D \in \mathcal{D}_0$. But as D runs over all positive operators in $\mathcal{D}_0$, so does D^2; by Proposition 2.6, this is enough to show $c_{\mathcal{D}}(R^2, S^2) = \|R^2\| \, \|S^2\|$.

4.2. THEOREM. R, S are optimally conditioned if and only if E_a, F_b are optimally conditioned for every $a, b < 1$.

Proof. Let R, S be given optimally conditioned. By Lemma 4.1, $\|R^pD^{-1}\| \cdot \|DS^p\| \geq 1$ for every positive $D \in \mathcal{D}_0$ and every $p \geq 1$ which is an integral power of 2. For fixed $a, b < 1$, we have $R^{2p} \leq E_a + a^{2p}I$ and $S^{2p} \leq F_b + b^{2p}I$; hence $\|R^pD^{-1}\|^2 = \|D^{-1}R^{2p}D^{-1}\| \leq \|E_aD^{-1}\|^2 + a^{2p}\|D^{-1}\|^2$; similarly $\|DS^p\|^2 \leq \|DF_b\|^2 + b^{2p}\|D\|^2$. Hold D fixed, and let $p \to \infty$ through integral powers of 2 to obtain $\|E_aD^{-1}\| \, \|DF_b\| \geq 1 = \|E_a\| \, \|F_b\|$. The pair E_a, F_b is therefore optimally conditioned.

Conversely, suppose R, S are given with E_a, F_b optimally conditioned for every $a, b < 1$. We have $R^2 \geq a^2E_a$, $S^2 \geq b^2F_b$, and so for each $D \in \mathcal{D}_0$, $\|RD^{-1}\|^2 \geq a^2\|E_aD^{-1}\|^2$, $\|DS\|^2 \geq b^2\|DF_b\|^2$. Since E_a, F_b are optimally conditioned, we have $\|RD^{-1}\| \, \|DS\| \geq ab\|E_aD^{-1}\| \, \|DF_b\| \geq ab\|E_a\| \, \|F_b\| = ab$. With $a, b \to 1$, we obtain $c_{\mathcal{D}}(R, S) \geq 1 = \|R\| \, \|S\|$.

4.3. COROLLARY. $\rho_{\mathcal{D},\mathcal{U}} = \sup\{[b_{\mathcal{U}}(E, F)]^{-1} : E, F$ self-adjoint projections such that $c_{\mathcal{D}}(E, F) = 1\}$.

Proof. $\rho_{\mathcal{D},\mathcal{U}} \geq \sup$ is immediate from the definition of $\rho_{\mathcal{D},\mathcal{U}}$. For the inequality in the reverse sense, we choose $\epsilon > 0$ and R, S positive of norm 1 such that $c_{\mathcal{D}}(R, S) = 1$ and $[b_{\mathcal{U}}(R, S)]^{-1} \geq \rho_{\mathcal{D},\mathcal{U}} - \epsilon$ ($b_{\mathcal{U}}(R, S) \leq \epsilon$ if $\rho_{\mathcal{D},\mathcal{U}} = \infty$), by Theorem 3.5. Let $a < 1$, and consider E_a, F_a. Since $a^2E_a \leq R^2$ and $a^2F_a \leq S^2$, we have for any $U \in \mathcal{U}$

$$\|RUS\|^2 = \|SU^*R^2US\| \geqslant a^2 \|SU^*E_aUS\| = a^2 \|E_aUS^2U^*E_a\| \geqslant a^4 \|E_aUF_b\|^2,$$

and so $b_{\mathcal{U}}(R, S) \geqslant a^2 b_{\mathcal{U}}(E_a, F_a)$. Hence $[b_{\mathcal{U}}(E_a, F_a)]^{-1} \geqslant a^2(\rho_{\mathcal{D},\mathcal{U}} - \epsilon)$ $([b_{\mathcal{U}}(E_a, F_a)]^{-1} \geqslant a^2\epsilon^{-1}$ if $\rho_{\mathcal{D},\mathcal{U}} = \infty)$ and now with $a \to 1$, then $\epsilon \to 0$, the proof is complete.

4.4. COROLLARY. Let Φ, Ψ be nondecreasing, nonnegative functions. Let R, S be positive and optimally conditioned. Then $\Phi(R), \Psi(S)$ are optimally conditioned.

Proof. The spectral resolutions of $\Phi(R), \Psi(S)$ are those of R, S respectively, and similarly ordered. R, S optimally conditioned implies, by the first half of Theorem 4.2, the optimal condition of their spectral resolutions. In turn, by the converse, this implies that $\Phi(R), \Psi(S)$ are optimally conditioned.

§5. COMMUTATIVE $\mathcal{D}$. In the case that $\mathcal{D}$ is commutative and $\mathcal{U}$ is the unitary group of $\mathcal{D}'$, we will show that $\rho_{\mathcal{D},\mathcal{U}} \leqslant 2$. This was proved in [1, Theorem 3] for the case of $\mathcal{H}$ finite dimensional and maximal abelian $\mathcal{D}$. Our approach is to follow the variational technique of [1] to prove $\rho_{\mathcal{D},\mathcal{U}} \leqslant 2$ generally in the case of finite dimensional $\mathcal{H}$, and then extend this bound to $\mathcal{H}$ of arbitrary dimension.

If $\mathcal{H}$ is of dimension $n < \infty$, we recall that a commutative von Neumann algebra can be represented, by choosing appropriate coordinates, as matricies in block diagonal form: $\mathrm{diag}(d_1 I_{n_1}, \ldots, d_k I_{n_k})$ where $n_1 + \ldots + n_k = n$ and I_{n_κ} is the $n_\kappa \times n_\kappa$ identity matrix. $\mathcal{U}$ is then represented by matricies in block diagonal form: $\mathrm{diag}(U_1, \ldots, U_k)$ where U_κ is an arbitrary $n_\kappa \times n_\kappa$ unitary matrix. We may think of vectors in $\mathcal{H}$ as k-tuples $x = (x_1, \ldots, x_k)$ where each x_κ is itself an n_κ-tuple. We partition the matrix representing R into $n \times n_\kappa$ blocks R_κ so that $R = \mathrm{row}(R_1, \ldots R_k)$, and we partition the matrix representing S into $n_\kappa \times n$ blocks S_κ so that $S = \mathrm{col}(S_1, \ldots, S_k)$. We then have $(x, RUSy) = \sum_{\kappa=1}^{k} (R_\kappa^* x, U_\kappa S_\kappa y)$.

5.1. LEMMA. Let $\mathcal{H}$ be finite dimensional, $\mathcal{D}$ abelian, and let $\mathcal{U}$ be the unitary group of $\mathcal{D}'$. Then $c_{\mathcal{D}}(R, S) \leqslant 2b_{\mathcal{U}}(R, S)$ for every R, S.

Proof. We adopt the variational procedure used in [1]. Let $U_0 \in \mathcal{U}$, and vectors u, v be chosen so that $(u, RU_0Sv) = b_{\mathcal{U}}(R, S)$. We note that $|(x, RUSy)| \leqslant b_{\mathcal{U}}(R, S)$ for every $U \in \mathcal{U}$ and unit vectors x, y; choosing U depending on x, y so that $(R_\kappa^* x, U_\kappa S_\kappa y) = |R_\kappa^* x| |S_\kappa y|$ for every κ, we see that

$$(5.1) \quad \sum_{\kappa=1}^{k} |R_\kappa^* x| |S_\kappa y| \leqslant b_{\mathcal{U}}(R, S), \quad \sum_{\kappa=1}^{k} |R_\kappa^* u| |S_\kappa v| = b_{\mathcal{U}}(R, S).$$

Assume for the moment that $|R_\kappa^* u| |S_\kappa v| = 0$ for every κ.

Let h be a unit vector orthogonal to u and set $x = \cos a\, u + \sin a\, h$ for small real a. Expanding (5.1) in powers of a, we have

$$(5.2)\quad \sum_{\kappa=1}^{k} |R_\kappa{}^*u|\,|S_\kappa v| + a\sum_{\kappa=1}^{k}(|S_\kappa v|/|R_\kappa{}^*u|)\mathrm{Re}(R_\kappa{}^*h, R_\kappa{}^*u) +$$
$$+ (a^2/2)\sum_{\kappa=1}^{k}[(|S_\kappa v|/|R_\kappa{}^*u|)|R_\kappa{}^*h|^2 - |S_\kappa v|\,|R_\kappa{}^*u| -$$
$$- (|\mathrm{Re}(R_\kappa{}^*h, R_\kappa{}^*v)|^2/|R_\kappa{}^*u|^3)|S_\kappa v|] + O(a^3) <$$
$$< b_{\mathcal{U}}(R,S) = \sum_{\kappa=1}^{k}|R_\kappa{}^*u|\,|S_\kappa v|.$$

The terms of order zero in a cancel. The term of first order in a yields, as $a \to 0$,

$$(5.3)\qquad (\mathrm{sgn}\, a)\sum_{\kappa=1}^{k}(|S_\kappa v|/|R_\kappa{}^*u|)\mathrm{Re}(R_\kappa{}^*h, R_\kappa{}^*u) \leqslant 0.$$

As a may be of either sign, and (5.3) is just as true for ih as h, we have in fact

$$\sum_{\kappa=1}^{k}(|S_\kappa v|/|R_\kappa{}^*u|)(R_\kappa{}^*h, R_\kappa{}^*u) = 0 \text{ for } h \perp u.$$

Thus u must be an eigenvector of $R\Delta^2R^*$, where $\Delta = \mathrm{diag}((|S_\kappa v|/|R_\kappa{}^*u|)^{1/2} I_{n_\kappa}) \in \in \mathcal{D}_0$; the corresponding eigenvalue is easily seen to be $b_{\mathcal{U}}(R,S)$. Turning to the terms of second order in a^2 in (5.2), we have

$$(5.4)\qquad (h, R\Delta^2R^*h) - b_{\mathcal{U}}(R,S) \leqslant$$
$$\leqslant \sum_{\kappa=1}^{k}(|S_\kappa v|/|R_\kappa{}^*u|^3)[\mathrm{Re}(R_\kappa{}^*h, R_\kappa{}^*u)]^2.$$

Writing (5.4) for ih also, adding (5.4) for h and ih, and using $|(R_\kappa{}^*h, R_\kappa{}^*u)|^2 \leqslant \leqslant |R_\kappa{}^*h|^2\,|R_\kappa{}^*u|^2$, we obtain

$$2(h, R\Delta^2R^*h) - 2b_{\mathcal{U}}(R,S) \leqslant (h, R\Delta^2R^*h).$$

This is true for any unit vector h orthogonal to u, and in particular if h is an eigenvector of $R\Delta^2R^*$ corresponding to any eigenvalue larger than $b_{\mathcal{U}}(R,S)$, so we have $\|R\Delta^2R^*\| \leqslant 2b_{\mathcal{U}}(R,S)$.

Holding u fixed and varying v leads similarly to $\|S^*D^2S\| \leqslant 2b_{\mathcal{U}}(R,S)$ where $D = \mathrm{diag}((|R_\kappa{}^*u|/|S_\kappa v|)^{1/2} I_{n_\kappa}) \in \mathcal{D}_0$ [perform the involution $(R,S,u,v) \to (S^*, R^*, v, u)$]. Noticing that $\Delta = D^{-1}$, we have

$$[c_{\mathcal{D}}(R,S)]^2 \leqslant \|RD^{-1}\|^2\,\|DS\|^2 = \|R\Delta^2R^*\|\,\|S^*D^2S\| \leqslant [2b_{\mathcal{U}}(R,S)]^2.$$

There remains only the onerous task of showing that there was no loss of generality in assuming that $|R_\kappa{}^*u|\,|S_\kappa v| = 0$ for every κ. By Theorem 3.5, there is no loss in generality in supposing initially that R and S are invertible. We will show that for every $\epsilon > 0$, there exist $R^{(\epsilon)}, S^{(\epsilon)}$ such that $\|R^{(\epsilon)} - R\| < \epsilon$, $\|S^{(\epsilon)} - S\| < < \epsilon$, and unit vectors $u^{(\epsilon)}, v^{(\epsilon)}$ with $|R^{(\epsilon)}{}_\kappa{}^*u^{(\epsilon)}|\,|S^{(\epsilon)}{}_\kappa v^{(\epsilon)}| = 0$ for every κ and such that $b_{\mathcal{U}}(R^{(\epsilon)}, S^{(\epsilon)}) = \sum_{\kappa=1}^{k} |R^{(\epsilon)}{}_{\kappa}{}^{*}u^{(\epsilon)}|\ |S^{(\epsilon)}{}_{\kappa}v^{(\epsilon)}|$; then we will finish with a continuity argument.

First we notice that $b_{\mathcal{U}}(R, S) = \sum_{\kappa=1}^{k} |R_{\kappa}{}^{*}u|\ |S_{\kappa}v|$ implies that for each κ, $|R_{\kappa}{}^{*}u| = 0$ if and only if $|S_{\kappa}v| = 0$. Indeed, suppose $R_1{}^{*}u = 0$. Let h be a unit vector such that $R_{\kappa}{}^{*}h = 0$ for $\kappa \geq 2$; since R is invertible, $R_1{}^{*}h = 0$. Multiply h by a scalar of modulus one so that $(u, h) \leq 0$ and thus $|\cos a\, u + \sin a\, h| \leq 1$ for small $a > 0$. Setting $x = \cos a\, u + \sin a\, h$, $y = v$ in (5.1), we have

$$\begin{aligned} b_{\mathcal{U}}(R, S) &\geq \sum_{\kappa=1}^{k} |R_{\kappa}{}^{*}x|\ |S_{\kappa}v| \\ &= a|R_1{}^{*}h|\ |S_1v| + \sum_{\kappa=2}^{k} |R_{\kappa}{}^{*}u|\ |S_{\kappa}v| + O(a^2) \\ &= a|R_1{}^{*}h|\ |S_1v| + b_{\mathcal{U}}(R, S) + O(a^2). \end{aligned}$$

As $a > 0$ and $|R_1{}^{*}h| = 0$, we obtain with $a \downarrow 0$ that $|S_1v| = 0$.

Next we observe that any neighborhood of R, S contains a pair R', S' such that $|(u', R'US'v')| = b_{\mathcal{U}}(R', S')$ for some U determines the unit vectors u', v' uniquely up to scalar multiples; put another way, if $U \in \mathcal{U}$ is such that $\|R'US'\| = b_{\mathcal{U}}(R', S')$, then for all such U the largest eigenvalues of $R'US'S'^{*}U^{*}R'^{*}$ and $S'^{*}U^{*}R'^{*}R'US'$ are simple, with eigenvectors u', v' respectively which do not depend on such U. Indeed, choose unit vectors u, v such that $|(u, RUSv)| = b_{\mathcal{U}}(R, S)$, let P be the operator which carries u into $R^{*}u$ and anihilates vectors orthogonal to u, and let Q be the operator which carries v into Sv and anihilates vectors orthogonal to v. Let $R' = R + \epsilon P^{*}$, $S' = S + \epsilon Q$ for $\epsilon > 0$ sufficiently small. We have trivially $b_{\mathcal{U}}(R', S') \leq (1 + \epsilon)^2 b_{\mathcal{U}}(R, S)$, $|(u, R'US'v)| = (1 + \epsilon)^2 |(u, RUSv)|$, so that $b_{\mathcal{U}}(R', S') = (1 + \epsilon)^2 b_{\mathcal{U}}(R, S)$ and $|(u, R'US'v)| = b_{\mathcal{U}}(R', S')$. For any unit vectors x, y, write $x = \cos a\, e^{i\theta}u + \sin a\, h$, $y = \cos\beta\, e^{i\phi}v + \sin\beta\, k$, with $h \perp u$, $k \perp v$, and $0 \leq a, \beta \leq \pi/2$. We obtain $|(x, R'US'y)| \leq (1 + \epsilon\cos a)(1 + \epsilon\cos\beta) b_{\mathcal{U}}(R, S)$, so that $|(x, R'US'y)| = b_{\mathcal{U}}(R', S')$ implies $\cos a = \cos\beta = 1$, and thus $u' = u$, $v' = v$ are as unique as claimed; further, $\|R'US'\| = b_{\mathcal{U}}(R', S')$ if and only if $|(u, RUSv)| = b_{\mathcal{U}}(R. S)$, and we denote the set of such U by $\mathcal{U}_0$. However, we need even more: we claim that there is a neighborhood of any such R', S' with the property that for all R'', S'' in this neighborhood, the unit vectors u'', v'' for which $|(u'', R''US''v'')| = b_{\mathcal{U}}(R'', S'')$ for some U are not only unique up to scalar multiples but may be taken close to u, v. To see this, observe that the simplicity of the largest eigenvalues of $R'US'S'^{*}U^{*}R'^{*}$ and $S'^{*}U^{*}R'^{*}R'US'$ for $U \in \mathcal{U}_0$, together with the compactness of $\mathcal{U}_0$, implies that there is an $\epsilon_0 > 0$ and an open neighborhood $\mathcal{V}$ of $\mathcal{U}_0$ in $\mathcal{U}$ such that for $\|R'' - R'\| < \epsilon_0$, $\|S'' - S'\| < \epsilon_0$, and $V \in \mathcal{V}$, the largest eigenvalues of $R''VS''S''^{*}V^{*}R''^{*}$ and $S''^{*}V^{*}R''^{*}R''VS''$ are simple and the corresponding unit eigenvectors u'', v'' can be chosen as continuous functions of R'', S'', V equal to u, v at R', S', and $V \in \mathcal{U}_0$. For $U \notin \mathcal{U}_0$ we have $\|R'US'\| < b_{\mathcal{U}}(R', S')$ and hence on the compact complement $\mathcal{W}$ of $\mathcal{V}$ in $\mathcal{U}$ we have for some $a > 0$, $\|R'WS'\| < b_{\mathcal{U}}(R', S') - a$ uniformly for $W \in \mathcal{W}$. Choose $\epsilon_1 \leq \epsilon_0$ such that $\|R'' - R'\| < \epsilon_1$, $\|S'' - S'\| < \epsilon_1$ imply

$\sup\{\|R''US'' - R'US'\|: U \in \mathcal{U}\} < a/2$. Then $b_{\mathcal{U}}(R'', S'') > b_{\mathcal{U}}(R', S') - a/2$, and for $W \in \mathcal{W}$ we have $\|R''WS''\| < \|R'WS'\| + a/2 < b_{\mathcal{U}}(R', S') - a/2 < b_{\mathcal{U}}(R'', S'')$. It follows that if the unit vectors u'', v'' and $V \in \mathcal{U}$ are such that $|(u'', R''VS''v'')| = b_{\mathcal{U}}(R'', S'')$, then $V \in \mathcal{V}$, and u'', v'' are unique up to scalar multiples; further, given any $\eta > 0$, we can find a positive $\epsilon_2 \leqslant \epsilon_1$ such that if $\|R'' - R'\| < \epsilon_2$, $\|S'' - S'\| < \epsilon_2$, we may assume that $|u'' - u| < \eta$ and $|v'' - v| < \eta$.

Now let J' be the set of indicies $\{\kappa: |R_\kappa'{}^* u'| = 0\} = \{\kappa: |S_\kappa' v| = 0\} = \{\kappa: |R_\kappa{}^* u| = |S_\kappa v| = 0\}$. We will show that in any neighborhood of R', S' there exist R'', S'' such that the corresponding set of indicies J'' is strictly contained in J'. By induction, choosing at each stage neighborhoods sufficiently small that $R^{(\mu)}, S^{(\mu)}$ can play the role of R', S', and then finding $R^{(\mu+1)}, S^{(\mu+1)}$ so that the corresponding set of indices $J^{(\mu+1)}$ is properly contained in $J^{(\mu)}$, we can, after a finite number of steps, find $R^{(m)}, S^{(m)}$ arbitrarily close to R', S' and hence arbitrarily close to R, S such that $J^{(m)}$ is empty. The construction of R'', S'' is as follows: suppose, for the sake of definiteness, that $1 \in J'$, $k \notin J'$. Define R'' by $R_1'' = R_1' + \epsilon R_k'$, $R_\kappa'' = R_\kappa'$ $(\kappa \geqslant 2)$, $S'' = S$. Choose $\epsilon > 0$ so small that the unit vectors u'', v'' are so close to u', v' that $|R_\kappa'{}^*u'|\,|S_\kappa' v'| \neq 0$ implies $|R_\kappa''{}^*u''|\,|S_\kappa'' v''| = |R_\kappa'{}^*u''|\,|S_\kappa' v''| \neq 0$ for every $\kappa \notin J'$. Thus $J'' \subset J'$. Now since $S_1'v' = S_1''v' = 0$, we have $\sum_{\kappa=1}^{k} |R_\kappa''{}^*u'|\,|S_\kappa'' v'| = \sum_{\kappa=2}^{k} |R_\kappa''{}^*u'|\,|S_\kappa' v'| = b_{\mathcal{U}}(R', S')$; but as $|S_1''v'| = 0$ while $|R_1''{}^*u'| \neq 0$, we cannot have $\sum_{\kappa=1}^{k} |R_\kappa''{}^*u'|\,|S_\kappa v'| = b_{\mathcal{U}}(R'', S'')$, and so $b_{\mathcal{U}}(R'', S'') > b_{\mathcal{U}}(R', S')$. This shows that the index 1, which was in J', cannot be in J'': $1 \in J''$ implies $S_1''v'' = 0$, which would give $b_{\mathcal{U}}(R'', S'') = \sum_{\kappa=2}^{k} |R_\kappa''{}^*u''|\,|S_\kappa'' v''| = \sum_{\kappa=2}^{k} |R_\kappa'{}^*u''|\,|S_\kappa' v''| \leqslant b_{\mathcal{U}}(R', S')$, a contradiction.

To complete the proof of this lemma, we choose for each $\epsilon > 0$, as we may, $R^{(\epsilon)}, S^{(\epsilon)}$ such that $\|R^{(\epsilon)} - R\| < \epsilon$, $\|S^{(\epsilon)} - S\| < \epsilon$, and $c_{\mathcal{D}}(R^{(\epsilon)}, S^{(\epsilon)}) \leqslant 2b_{\mathcal{U}}(R^{(\epsilon)}, S^{(\epsilon)})$. Recall that R, S were initially chosen invertible without loss of generality, so we can find $D^{(\epsilon)} \in \mathcal{D}_0$ such that $c_{\mathcal{D}}(R^{(\epsilon)}; S^{(\epsilon)}) = \|R^{(\epsilon)} D^{(\epsilon)-1}\|\,\|D^{(\epsilon)} S^{(\epsilon)}\|$ and constants $0 < m, M < \infty$ for which $m \leqslant D^{(\epsilon)} \leqslant M$ uniformly for small ϵ (Lemma 3.1 and Theorem 3.3). Let $\epsilon \downarrow 0$ through a sequence of values such that $D^{(\epsilon)}$ converges to some D. Then as $D^{(\epsilon)-1}$ converges to D^{-1} also, we have

$$c_{\mathcal{D}}(R, S) \leqslant \|RD^{-1}\|\,\|DS\| = \lim_{\epsilon\to 0} \|R^{(\epsilon)} D^{(\epsilon)-1}\|\,\|D^{(\epsilon)} S^{(\epsilon)}\|$$

$$= \lim_{\epsilon\to 0} c_{\mathcal{D}}(R^{(\epsilon)}, S^{(\epsilon)}) \leqslant 2 \lim_{\epsilon\to 0} b_{\mathcal{U}}(R^{(\epsilon)}, S^{(\epsilon)}) = 2b_{\mathcal{U}}(R, S).$$

Having completed the estimate for finite dimensional $\mathcal{H}$, we now turn to the case of infinite dimensional $\mathcal{H}$.

5.2. THEOREM. If $\mathcal{D}$ is commutative and $\mathcal{U}$ is the unitary group of $\mathcal{D}'$, then $\rho_{\mathcal{D},\mathcal{U}} \leqslant 2$.

Proof. (a) We first consider the case where $\mathcal{H}$ is separable and $\mathcal{D}$ is of finite multiplicity. In this case we may represent $\mathcal{H}$ as $\sum_{\nu=1}^{n} \oplus \mathcal{L}_2([0, 1], d\mu_\nu)$ with μ_ν

nonnegative Borel measures, so that $x \in \mathcal{H}$ is represented as an n-tuple of Borel functions $x = \{x_\nu(\omega)\}$ with norm given by $|x|^2 = \sum_{\nu=1}^{n} \int_0^1 |x_\nu(\omega)|^2 \mu_\nu(d\omega)$, and $D \in \mathcal{D}$ effects the mapping $\{x_\nu(\omega)\} \to \{d(\omega)x_\nu(\omega)\}$ with $d(\omega)$ a bounded Borel function. $U \in \mathcal{U}$ is then represented by an $n \times n$ unitary matrix valued function $(u_{\mu\nu}(\omega))$. Let π denote a partition of $[0, 1]$ into a finite number of Borel sets, write $\sigma > \pi$ if σ refines π, and denote by E_π the self-adjoint projection of $\mathcal{H}$ onto the finite dimensional subspace consisting of functions which are measurable with respect to the field of sets generated by π. Fix R, S positive and invertible, and set $R_\pi = E_\pi R E_\pi$, $S_\pi = E_\pi S E_\pi$, let $\mathcal{U}_\pi$ be the subgroup of $\mathcal{U}$ consisting of those $(u_{\mu\nu}(\omega))$ which are constant on the sets of π, and let $\mathcal{D}_\pi$ be the subalgebra of $\mathcal{D}$ consisting of those $d(\omega)$ which are constant on the sets of π. In the finite dimensional space $\mathcal{H}_\pi = E_\pi \mathcal{H}$, $\mathcal{D}_\pi$ is an abelian von Neumann algebra and the unitary group of its commutant is $\mathcal{U}_\pi$. R_π and S_π, now considered as operators on $\mathcal{H}_\pi$, are not only self-adjoint, but have upper bounds in the sense of positive definiteness between those of R and S: the bounds of R, S are extrema over all unit vectors in $\mathcal{H}$, while the bounds of R_π, S_π are only extrema over unit vectors in $\mathcal{H}_\pi$. Thus by Theorem 3.3 there exists a positive $D_\pi \in \mathcal{D}_\pi$ for which $c_{\mathcal{D}_\pi}(R_\pi, S_\pi) = \|R_\pi D_\pi^{-1}\| \, \|D_\pi S_\pi\|$; by Lemma 5.1, $c_{\mathcal{D}_\pi}(R_\pi, S_\pi)$ is dominated by $2b_{\mathcal{U}_\pi}(R_\pi, S_\pi)$. Further, we may take D_π so that $\|R_\pi D_\pi^{-1}\| = \|D_\pi S\pi\|$ and $m \leqslant D \leqslant M$ uniformly in π by Lemma 3.1, where m, M are the constants given for R, S in that lemma.

Now we let $\mathcal{K}_\pi$ be the convex hull of the operators $\{D_\sigma^2 : \sigma > \pi\}$, closed in the weak, hence also in the strong, operator topology. Each $\mathcal{K}_\pi$ is compact in the weak operator topology, and the family of sets $\mathcal{K}_\pi$ has the finite intersection property; we may therefore choose D_0 positive so that $D_0^2 \in \bigcap_\pi \mathcal{K}_\pi$. We have, just as in the proof of Theorem 3.3, that $\|E_\pi S D_0{}^2 S E_\pi\| \leqslant \sup_{\sigma > \pi} c_{\mathcal{D}_\sigma}(R_\sigma, S_\sigma)$ and $\|E_\pi R D_0{}^{-2} R E_\pi\| \leqslant \sup_{\sigma > \pi} c_{\mathcal{D}_\sigma}(R_\sigma, S_\sigma)$. Noting that $E_\pi R \to R$, $E_\pi S \to S$ strongly, we have $\|R D_0^{-1}\| \cdot \|DS\| \leqslant \sup_\pi c_{\mathcal{D}_\pi}(R_\pi, S_\pi)$. Finally, for every π we have $b_{\mathcal{U}_\pi}(R_\pi, S_\pi) \leqslant b_{\mathcal{U}}(R_\pi, S_\pi)$. For every $U \in \mathcal{U}$ we have $\|R_\pi U S_\pi\| \leqslant \|R E_\pi U E_\pi S\|$; $E_\pi U E_\pi$ is an operator of norm 1 in $\mathcal{D}'_\pi$ and hence is a convex combination of unitaries in $\mathcal{D}'_\pi \subset \mathcal{D}'$ (remembering again that $\mathcal{H}_\pi$ is finite dimensional so that every contraction on $\mathcal{H}_\pi$ is in the convex hull of the unitaries) and it follows that $b_{\mathcal{U}_\pi}(R_\pi, S_\pi) \leqslant b_{\mathcal{U}}(R_\pi, S_\pi) \leqslant b_{\mathcal{U}}(R, S)$ for every π. The estimate $c_{\mathcal{D}_\pi}(R_\pi, S_\pi) \leqslant 2b_{\mathcal{U}_\pi}(R_\pi, S_\pi)$ thus gives $\|R D_0{}^{-1}\| \, \|D_0 S\| \leqslant 2b_{\mathcal{U}}(R, S)$ as desired.

(b) If $\mathcal{H}$ is inseparable and/or $\mathcal{D}$ is of infinite multiplicity, we represent $\mathcal{H}$ as $\sum_{a \in A} \oplus \mathcal{L}_2(\Omega, d\mu_a)$ where Ω is the Stone space of the lattice of self-adjoint projections in $\mathcal{D}$, $d\mu_a$ are nonnegative regular Borel measures on Ω, and the operators in $\mathcal{D}$ are represented as $D\{x_a\} = \{d(\omega)x_a(\omega)\}$ for continuous $d(\omega)$; the index set A need not be countable nor need the spaces $\mathcal{L}_2(\Omega, d\mu_a)$ be separable. We consider families $\mathcal{P} = \{\{\pi_a\} : \pi_a \subset \Omega$, $\pi_a = 0$ except for a finite number of a, and $\mathcal{L}_2(\pi_a, d\mu_a)$ is separable for every $a\}$. Let $E_{\mathcal{P}}$ be the self-adjoint projection on $\mathcal{H}$ with range the closure of $\{\{x_a(\omega)\} : x_a(\omega)$ has support in $\pi_a\}$; $\mathcal{Q} \geqslant \mathcal{P}$ means $E_{\mathcal{Q}} \geqslant E_{\mathcal{P}}$.

We denote by $R_{\mathcal{P}}, S_{\mathcal{P}}$ the operators $E_{\mathcal{P}}RE_{\mathcal{P}}, E_{\mathcal{P}}SE_{\mathcal{P}}$, by $\mathcal{D}_{\mathcal{P}}$ the von Neumann algebra $E_{\mathcal{P}}\mathcal{D}E_{\mathcal{P}}$ on $E_{\mathcal{P}}\mathcal{H}$, and by $\mathcal{U}_{\mathcal{P}}$ the unitary group of the commutant of $\mathcal{D}_{\mathcal{P}}$; $\mathcal{D}_{\mathcal{P}}$ may be thought of as imbedded in $\mathcal{D}$ by taking $d(\omega) = 1$ off $\cup_a \pi_a$. Since the upper and lower bounds of $R_{\mathcal{P}}$ and $S_{\mathcal{P}}$ are contained between those of R and S, we know that there exist $D_{\mathcal{P}} \in \mathcal{D}_{\mathcal{P}}$ such that $m \leqslant$
Lemma 3.1 for R, S) with $\|R_{\mathcal{P}}D_{\mathcal{P}}^{-1}\|^2 = \|D_{\mathcal{P}}S_{\mathcal{P}}\|^2 = c_{\mathcal{D}_{\mathcal{P}}}(R_{\mathcal{P}}, S_{\mathcal{P}}) \leqslant 2b_{\mathcal{U}_{\mathcal{P}}}(R_{\mathcal{P}}, S_{\mathcal{P}})$; the las inequality comes from (a) above. Let $\mathcal{K}_{\mathcal{P}}$ be the weak operator closure of the convex hull of all $D_{\mathcal{Q}}^2$ for $\mathcal{Q} \geqslant \mathcal{P}$, so that the sets $\mathcal{K}_{\mathcal{P}}$ are convex, compact, closed in the strong operator topology, and have the finite intersection property. As before using the fact that $E_{\mathcal{P}}$ converges to the identity in the strong operator topology and that norms are lower semi-continuous in this topology, we obtain for $D_0^2 \in \cap_{\mathcal{P}}\mathcal{K}_{\mathcal{P}}$ that $\|RD_0^{-1}\|\,\|D_0S\| \leqslant \lim_{\mathcal{P}} \sup_{\mathcal{Q}>\mathcal{P}} 2b_{\mathcal{U}_{\mathcal{Q}}}(R_{\mathcal{Q}}, S_{\mathcal{Q}})$. Now for any $\mathcal{Q}$ we have $b_{\mathcal{U}_{\mathcal{Q}}}(R_{\mathcal{Q}}, S_{\mathcal{Q}}) \leqslant b_{\mathcal{U}}(R, S)$; for, given $U_{\mathcal{Q}} \in \mathcal{U}_{\mathcal{Q}}$, we have

$$\|R_{\mathcal{Q}}U_{\mathcal{Q}}S_{\mathcal{Q}}\| = \|E_{\mathcal{Q}}RE_{\mathcal{Q}}U_{\mathcal{Q}}E_{\mathcal{Q}}SE_{\mathcal{Q}}\| \leqslant \|RE_{\mathcal{Q}}U_{\mathcal{Q}}E_{\mathcal{Q}}S\|,$$

and $E_{\mathcal{Q}}U_{\mathcal{Q}}E_{\mathcal{Q}}$ is the average of $U_{\mathcal{Q}} + (I - E_{\mathcal{Q}}) = U \in \mathcal{U}$ and $U_{\mathcal{Q}} - (I - E_{\mathcal{Q}}) = U' \in \in \mathcal{U}$, so $\|RE_{\mathcal{Q}}U_{\mathcal{Q}}E_{\mathcal{Q}}S\| \leqslant \frac{1}{2}\|RUS\| + \frac{1}{2}\|RU'S\| \leqslant b_{\mathcal{U}}(R, S)$. It follows that $c_{\mathcal{D}}(R, S) < \|RD_0^{-1}\|\,\|D_0S\| \leqslant 2b_{\mathcal{U}}(R, S)$ as desired.

§6. AN EXPLICIT TEST FOR OPTIMAL CONDITION. Throughout this section we will assume that the dimension of $\mathcal{H}$ is finite, $\dim \mathcal{H} = n$. We may therefore assume that $\mathcal{D}$ is given as the matrix algebra $\mathcal{D} = \sum_{\kappa=1}^{k} \oplus \mathcal{M}_{m_\kappa} I_{n_\kappa}$ where $\mathcal{M}_{m_\kappa}$ is the set of all $m_\kappa \times m_\kappa$ matrices, I_{n_κ} is the $n_\kappa \times n_\kappa$ identity matrix, and $m_1n_1 + \ldots + m_kn_k = n$; $D \in \mathcal{D}$ if and only if $D = \text{diag}(D_1, \ldots, D_1; D_2, \ldots, D_2; \ldots; D_k, \ldots, D_k)$ where each D_κ is an arbitrary element of $\mathcal{M}_{m_\kappa}$ and there are n_κ occurrences of each D_κ [2; Theorem 7, p. 8]. For an arbitrary matrix A, we perform the coherent partitioning and denote by $A_{\kappa,\mu;\lambda,\nu}$ $(1 \leqslant \mu \leqslant n_\kappa,\ 1 \leqslant \nu \leqslant n_\lambda)$ the block corresponding to the μ-th occurrence of $\mathcal{M}_{m_\kappa}$ and the ν-th occurrence of $\mathcal{M}_{m_\lambda}$:

$$A = \begin{pmatrix} A_{1,1;1,1} & A_{1,1;1,2} & \cdots A_{1,1;1,n_1} & A_{1,1;2,1} & \cdots & \\ A_{1,2;1,1} & A_{1,2;1,2} & \cdots & & & \\ & \cdots & & & & \\ A_{1,n_1;1,1} & \cdots & A_{1,n_1;1,n_1} & A_{1,n_1;2,1} & \cdots & \\ A_{2,1;1,1} & \cdots & A_{2,1;1,n_1} & A_{2,1;2,1} & & \\ & \cdots & & & & \\ & & & & & A_{k,n_k;k,n_k} \end{pmatrix}$$

For matrices we introduce the norm $\|\cdot\|_p$: $\|A\|_p = [\text{trace}\,(A^*A)^{p/2}]^{1/p}$; $\|A\| \leqslant \leqslant \|A\|_p \leqslant n^{1/p}\|A\|$ for every A, and so we see that as $p \to \infty$, $\|A\|_p \to \|A\|$. Further, the norm $\|\cdot\|_p$ is uniformly, hence strictly convex ($\|A + B\|_p = \|A\|_p + \|B\|_p$, $A = 0$, implies $B = cA$ for some $c \geqslant 0$) [3, Theorem 2.7]. Given R, S and $p < \infty$, we define $\Phi(\Delta) = \|R\Delta^{-1}R^*\|_p^{1/2}\|S^*\Delta S\|_p^{1/2}$ for positive $\Delta \in \mathcal{D}_0$.

6.1. LEMMA. (i) Φ is a convex function of Δ.

(ii) If R or S is invertible, and Δ, Δ' are such that $\Phi((1-t)\Delta + t\Delta') = (1-t)\Phi(\Delta) + t\Phi(\Delta')$ for some t between 0 and 1, then $\Delta' = c\Delta$ for some $c > 0$.

(iii) If R and S are both invertible, then Φ attains its minimum; if then Δ and Δ' are such that $\Phi(\Delta) = \Phi(\Delta') = \min \Phi$, then $\Delta' = c\Delta$ for some $c > 0$.

(iv) If p is integral, then the only stationary points for Φ can be where Φ attains its minimum.

Proof. (i) Given Δ, Δ' we set $a = \|R\Delta^{-1}R^*\|_p^{1/2}\|S^*\Delta S\|_p^{-1/2}$, $a' = \|R\Delta'^{-1}R^*\|_p^{1/2} \cdot \|S^*\Delta' S\|^{-1/2}$ so that $\Phi(\Delta) = \|R(a\Delta)^{-1}R^*\|_p = \|S^*(a\Delta)S\|_p$ and $\Phi(\Delta') = \|R(a'\Delta')^{-1} \cdot R^*\|_p = \|S^*(a'\Delta')S\|_p$. Since $(1-t)\Delta + t\Delta' = b[(1-\tau)a\Delta + \tau a'\Delta']$ with $\tau = at/[(1-t)a' + ta]$ and $b = [(1-t)a' + ta]/aa'$, and since $\Phi(c\Delta) = \Phi(\Delta)$ for any $c > 0$ and any Δ, it is no loss of generality to suppose that $a = a' = 1$. We then have for $0 < t < 1$

$$(6.1) \qquad \|S^*[(1-t)\Delta + t\Delta']S\|_p \leqslant (1-t)\|S^*\Delta S\|_p + t\|S^*\Delta' S\|_p = (1-t)\Phi(\Delta) + t\Phi(\Delta'),$$

with equality only if $S^*\Delta' S = cS^*\Delta S$ for some $c > 0$. Similarly

$$(6.2) \qquad \begin{aligned} \|R[(1-t)\Delta + t\Delta']^{-1}R^*\|_p &\leqslant \|R[(1-t)\Delta^{-1} + t\Delta'^{-1}]R^*\|_p \\ &\leqslant (1-t)\|R\Delta^{-1}R^*\|_p + t\|R\Delta'^{-1}R^*\|_p \\ &= (1-t)\Phi(\Delta) + t\Phi(\Delta') \end{aligned}$$

with equality in the second inequality only if $R\Delta'^{-1}R^* = cR\Delta^{-1}R^*$ for some $c > 0$; the first inequality follows from Lemma 3.2: $[(1-t)\Delta + t\Delta']^{-1} \leqslant (1-t)\Delta^{-1} + t\Delta'^{-1}$, which implies $R[(1-t)\Delta + t\Delta']^{-1}R^* \leqslant R[(1-t)\Delta^{-1} + t\Delta^{-1}]R^*$. The product of (6.1) and (6.2) yields the convexity of Φ:

$$[\Phi((1-t)\Delta + t\Delta')]^2 \leqslant [(1-t)\Phi(\Delta) + t\Phi(\Delta')]^2$$

(ii) To have $\Phi((1-t)\Delta + t\Delta') = (1-t)\Phi(\Delta) + t\Phi(\Delta')$ for some t between 0 and 1 requires equality in both (6.1) and (6.2). If S is invertible, we may infer from $S^*\Delta' S = cS^*\Delta S$ that $\Delta' = c\Delta$; if R is invertible, we may infer the same from $R\Delta'^{-1}R^* = cR\Delta^{-1}R^*$.

(iii) That Φ attains its minimum follows from first an *a priori* estimate like that of Lemma 3.1, and then since we are in a finite dimensional situation we may simply take any limit of a minimizing sequence of Δ as a point at which Φ is minimized. To see that the minimum is now unique up to scalar multiples, suppose $\Phi(\Delta) = \Phi(\Delta') = \min \Phi$. Then

$$\min \Phi \leqslant \Phi(\tfrac{1}{2}(\Delta + \Delta')) \leqslant \text{(by (i))}\ \tfrac{1}{2}\Phi(\Delta) + \tfrac{1}{2}\Phi(\Delta') = \min \Phi$$

and so $\Phi(\frac{1}{2}(\Delta + \Delta')) = \frac{1}{2}\Phi(\Delta) + \frac{1}{2}\Phi(\Delta')$. By (ii), $\Delta = c\Delta'$ for some $c > 0$.

(iv) If p is integral, then $\Phi(\Delta)^p = \operatorname{trace}(R\Delta^{-1}R^*)^p \cdot \operatorname{trace}(S^*\Delta S)^p$ is a polynomial in the real and imaginary parts of the entries of Δ and Δ^{-1}, and hence is a rational function in the entries of Δ; hence $\Phi(\Delta)$ must be a real analytic function of the real and imaginary parts of the entries of Δ. Clearly a minimum is a stationary point of Φ. Conversely, suppose Δ is not a minimum of Φ. Choose Δ' so that $\Phi(\Delta') < \Phi(\Delta)$, and consider $\frac{d}{dt}\Phi((1-t)\Delta + t\Delta')_{t=0+}$. We have

$$\begin{aligned}\Phi((1-t)\Delta + t\Delta') - \Phi(\Delta) &\leqslant [(1-t)\Phi(\Delta) + t\Phi(\Delta')] - \Phi(\Delta) \\ &= t[\Phi(\Delta') - \Phi(\Delta)],\end{aligned}$$

and hence $\frac{d}{dt}\Phi((1-t)\Delta + t\Delta')_{t=0+} \leqslant \Phi(\Delta') - \Phi(\Delta) < 0$. Therefore Δ cannot be a stationary point for Φ.

6.2. LEMMA. Let R, S be given positive semi-definite, and p integral. Then I is a stationary point for Φ if and only if there exists $c > 0$ such that for every κ

$$\sum_{\nu=1}^{n_\kappa} (R^{2p})_{\kappa,\nu;\kappa,\nu} = c \sum_{\nu=1}^{n_\kappa} (S^{2p})_{\kappa,\nu;\kappa,\nu}.$$

Proof. Set $H = \operatorname{diag}(H_1, \ldots, H_1; H_2, \ldots, H_2; \ldots; H_k, \ldots, H_k)$ where there are n_κ occurrences of the $m_\kappa \times m_\kappa$ Hermitian matrix H_κ. For small real ϵ, $I + \epsilon H$ is positive and belongs to $\mathcal{D}$. We write

$$\frac{\partial\Phi^p}{\partial\Delta}(I; H) = \lim \frac{1}{\epsilon}[\Phi^p(I + \epsilon H) - \Phi^p(I)],$$

the directional derivative of Φ^p at I in the direction H. We have

$$\begin{aligned}\frac{\partial\Phi^p}{\partial\Delta}(I; H) = \sum_{\kappa=1}^{k}\{&\operatorname{trace}[\sum_{q=0}^{p-1} R^{2q}(-RH_\kappa R)R^{2(p-1-q)}] \cdot \operatorname{trace}(S^{2p}) + \\ &+ \operatorname{trace}(R^{2p}) \cdot \operatorname{trace}[\sum_{q=0}^{p-1} S^{2q}(SH_\kappa S)S^{2(p-1-q)}]\}\end{aligned}$$

where we have confused H_κ and $\operatorname{diag}(0, \ldots, 0; \ldots; H_\kappa, \ldots, H_\kappa; 0, \ldots, 0; \ldots)$. Now $\operatorname{trace}[R^{2q}(-RH_\kappa)R^{2(p-1-q)}] = -\operatorname{trace}(H_\kappa R^{2p}) = -\operatorname{trace}(H_\kappa \sum_{\nu=1}^{n_\kappa} (R^{2p})_{\kappa,\nu;\kappa,\nu})$, and similarly $\operatorname{trace}[S^{2q}(SH_\kappa S)S^{2(p-1-q)}] = \operatorname{trace}(H_\kappa \sum_{\nu=1}^{n_\kappa} (S^{2p})_{\kappa,\nu;\kappa,\nu})$.

$$\frac{\partial\Phi^p}{\partial\Delta}(I;H) = p\sum_{\kappa=1}^{k}\Big[-\operatorname{trace}\Big(H_\kappa\sum_{\nu=1}^{n_\kappa}(R^{2p})_{\kappa,\nu;\kappa,\nu}\Big)\cdot\operatorname{trace}(S^{2p}) +$$
$$+\operatorname{trace}(R^{2p})\cdot\operatorname{trace}\Big(H_\kappa\sum_{\nu=1}^{n_\kappa}(S^{2p})_{\kappa,\nu;\kappa,\nu}\Big)\Big].$$

Our conditions on R and S show that $\operatorname{trace}(H_\kappa \sum_{\nu=1}^{n_\kappa}(R^{2p})_{\kappa,\nu;\kappa,\nu}) = c\operatorname{trace}(H_\kappa \sum_{\nu=1}^{n_\kappa}(S^{2p})_{\kappa,\nu;\kappa,\nu})$ for every κ and H_κ, and also that $\operatorname{trace}(R^{2p}) = c\operatorname{trace}(S^{2p})$. Our conditions on R and S therefore yield $\frac{\partial\Phi^p}{\partial\Delta}(I;H) = 0$ for every H, so I must be a stationary point for Φ.

Conversely, if I is a stationary point for Φ, we have $\frac{\partial\Phi^p}{\partial\Delta}(I, H_\kappa) = 0$ for every H, which yields for every κ

$$\frac{\operatorname{trace}(H_\kappa \sum_{\nu=1}^{n_\kappa}(R^{2p})_{\kappa,\nu;\kappa,\nu})}{\operatorname{trace}(H_\kappa \sum_{\nu=1}^{n_\kappa}(S^{2p})_{\kappa,\nu;\kappa,\nu})} = \frac{\operatorname{trace}(R^{2p})}{\operatorname{trace}(S^{2p})} = c.$$

We therefore need only prove generally that if A and B are matrices such that trace (KA) = trace(KB) for every Hermitian K, then A = B; for then taking $K = H_\kappa$, $A = \sum_{\nu=1}^{n}(R^{2p})_{\kappa,\nu;\kappa,\nu}$, $B = \sum_{\nu=1}^{n}(S^{2p})_{\kappa,\nu;\kappa,\nu}$, we will be done. Writing $A = (a_{\mu\nu})$, $B = (b_{\mu\nu})$, $K = (k_{\mu\nu})$, we see first that by taking $k_{\mu\nu} = \delta_{\kappa\mu}\delta_{\kappa\nu}$ that trace(KA) = = trace(KB) implies $a_{\kappa\kappa} = b_{\kappa\kappa}$ for every κ. By taking $k_{\mu\nu} = \delta_{\kappa\mu}\delta_{\lambda\nu} + \delta_{\lambda\mu}\delta_{\kappa\nu}$ we see that $a_{\kappa\lambda} + a_{\lambda\kappa} = b_{\kappa\lambda} + b_{\lambda\kappa}$; by taking $k_{\mu\nu} = i\delta_{\kappa\mu}\delta_{\lambda\nu} - i\delta_{\lambda\mu}\delta_{\kappa\nu}$, we see that $-ia_{\kappa\lambda} + ia_{\lambda\kappa} = -ib_{\kappa\lambda} + ib_{\lambda\kappa}$. It follows that $a_{\kappa\lambda} = b_{\kappa\lambda}$ for every κ, λ.

6.3. THEOREM. Let E, F be two self-adjoint projections with the property that there exists a number c for which $\sum_{\nu=1}^{n_\kappa} E_{\kappa,\nu;\kappa,\nu} = c\sum_{\nu=1}^{n_\kappa} F_{\kappa,\nu;\kappa,\nu}$ for every κ. Then $c_{\mathcal{D}}(E, F) = \|E\|\,\|F\| = 1$, and E, F are optimally conditioned with respect to $\mathcal{D}$.

Proof. From Lemma 6.2 and the fact that $E^{2p} = E$, $F^{2p} = F$, we see that the function $\Phi_p(\Delta) = \|E\Delta^{-1}E\|_p\,\|F\Delta F\|_p$ has a stationary point at $\Delta = I$ for every integral p. By Lemma 6.1 (iv), $\Phi_p(\Delta) \geqslant \Phi_p(I)$ for every positive $\Delta \in \mathcal{D}$.

Now suppose that there were a $D \in \mathcal{D}$ such that $\|ED^{-1}\|\,\|DF\| < 1$. Choose p so large that $\|ED^{-1}\|_{2p}^2\,\|DF\|_{2p}^2 < 1 \leqslant \|E\|_{2p}^2\,\|F\|_{2p}^2$. For this p we would have $\Phi_p((D^*D)^{1/2}) = \|ED^{-1}\|_{2p}^2\,\|DF\|_{2p}^2 < \|E\|_{2p}^2\,\|F\|_{2p}^2 = \Phi_p(I)$, a contradiction.

Given arbitrary R, S, we see that a sufficient condition that R, S be optimally conditioned with respect to $\mathcal{D}$ is that the projections E, F associated with the largest eigenvalues of R*R and SS* satisfy the hypothesis of Theorem 6.3; Proposition 2.5 allows us to consider $(R^*R)^{1/2}$, $(SS^*)^{1/2}$ in place of R, S, and then Theorem 4.2 shows that it is only the behavior of E, F that matters.

We wish to point out that the condition of Theorem 6.3 implies that E, F are optimally conditioned using the techniques of Forsythe and Straus [4]; many of the results of §4 may also be obtained using their methods.

The converse to Theorem 6.3 is not true: let $\mathcal{D}$ be the set of 2×2 diagonal matrices, $E = \begin{pmatrix} 1 & 0 \\ 0 & 0 \end{pmatrix}$, $F = \begin{pmatrix} 1 & 0 \\ 0 & 1 \end{pmatrix}$. We conjectture, however, that in general if E, F are self-adjoint projections with $c_{\mathcal{D}}(E, F) = 1$, then there exist self-adjoint projections $P \leqslant E$, $Q \leqslant F$ such that P, Q satisfy the hypothesis of Theorem 6.3; but we do not even know if $\rho_{\mathcal{D},\mathcal{U}}$ is determined by those pairs of self-adjoint projections E, F which satisfy the hypotheses of Theorem 6.3.

With the help of Theorem 6.3, it is possible to show that in general $\rho_{\mathcal{D},\mathcal{U}}$ need not be finite if $\mathcal{D}$ is not commutative, even if $\mathcal{U}$ is the full unitary group of $\mathcal{D}'$.

6.4. EXAMPLE. Consider the four 2×2 matrices $H_0 = \begin{pmatrix} 1 & 0 \\ 0 & 1 \end{pmatrix}$, $H_1 = \begin{pmatrix} 1 & 0 \\ 0 & -1 \end{pmatrix}$, $H_2 = \begin{pmatrix} 0 & 1 \\ 1 & 0 \end{pmatrix}$, $H_3 = \begin{pmatrix} 0 & i \\ -i & 0 \end{pmatrix}$; these are orthogonal and of norm $\sqrt{2}$ in the Hilbert-Schmidt norm, so if $A = \sum_{\nu=0}^{3} \omega_\nu H_\nu$ $(|\omega_\nu| = 1)$, then $\|A\| \leqslant \|A\|_2 = \sqrt{8}$. Fix an integer k and consider the 4^k matrices $H_a = H_{a_0} \otimes \ldots \otimes H_{a_{k-1}}$ of size $2^k \times 2^k$ where $a = (a_0, \ldots, a_{k-1})$ is a k-multi-index with each a_κ equal to $0, 1, 2$, or 3; for bookkeeping purposes, we set $|a| = \sum_{\kappa=0}^{k-1} 4^\kappa a_\kappa + 1$. Each H_a is a self-adjoint unitary, and they are all orthogonal and of norm $2^{k/2}$ in the Hilbert-Schmidt norm: $(H_a, H_\beta) = \text{trace} H_a \cdot H_\beta^* = \prod_{\kappa=0}^{k-1} \text{trace}(H_{a_\kappa} \cdot H_{\beta_\kappa}{}^*) = 2^{k/2} \delta_{a\beta}$. Thus if $X = \sum_{|a|=1}^{4^k} \omega_{|a|} H_a$, $\omega_{|\alpha|} = 1$ for all a, we have $\|X\| \leqslant \|X\|_2 \leqslant (4^k)^{1/2} \cdot \sup_a \|H_a\|_2 = 8^{k/2}$.

For convenience, set $n = 2^k$. Let $\mathcal{H}$ be of dimension n^3 and let E be the matrix given in $n^2 \times n^2$ blocks each of size $n \times n$:

$$E = \frac{1}{n^2} \begin{pmatrix} I & I & \ldots & I \\ I & I & \ldots & I \\ & \ldots & & \\ I & I & \ldots & I \end{pmatrix}, \quad I \text{ the } n \times n \text{ identity matrix.}$$

Let W denote the $n^3 \times n^3$ self-adjoint unitary $\text{diag}(H_1, \ldots, H_{n^2})$ presented in block diagonal form, where we have written $H_{|a|}$ for the $n \times n$ self-adjoint unitary denoted H_a above. Let $F = WEW^*$. Note that the diagonal $n \times n$ blocks of E and F are the same and that E and F are self-adjoint projections; by Theorem 6.3, E, F are optimally conditioned with respect to the von Neumann algebra $\mathcal{D} = \{\text{diag}(D_1, \ldots, D_{n^2})$: D arbitrary $n \times n$ matrices$\}$. The unitary group of $\mathcal{D}'$ is $\mathcal{U} = \{\text{diag}(\omega_1 I, \ldots, \omega_{n^2} I)$: $|\omega_\nu| = 1$, I the $n \times n$ identity$\}$.

Let us calculate $b_{\mathcal{U}}(E, F)$. We have $\|EUF\| = \|EUWE\|$; $EUWE$ is of the form

$$EUWE = \frac{1}{n^2} \begin{pmatrix} X & X & \ldots & X \\ X & X & \ldots & X \\ & \ldots & & \\ X & X & \ldots & X \end{pmatrix}$$

where $X = \sum_{\nu=1}^{n^2} \omega_\nu H_\nu$ with $|\omega_\nu| = 1$. Thus $\|EUWE\| = n^{-2}\|X\| \leqslant n^{-2}\|X\|_2 = n^{-1/2}$

for every $U \in \mathcal{U}$. Hence $b_{\mathcal{U}}(E, F) \leq n^{-\frac{1}{2}}$ and $\rho_{\mathcal{U}}(E, F) \geq n^{\frac{1}{2}}$.

If we take the direct sum of these examples for an unbounded sequence of k, we obtain an example with $\rho_{\mathcal{D},\mathcal{U}} = \infty$.

§7. OTHER TYPES OF RESULTS. First we wish to discuss an estimate of a related but different nature.

7.1. THEOREM. Suppose that $\mathcal{D}$ is finite dimensional and that $\mathcal{U}$ is any unitary subgroup of $\mathcal{D}'$ with the property that $AU = UA$ for all $U \in \mathcal{U}$ implies $A \in \mathcal{D}$. Then for invertible S, $c_{\mathcal{D}}(S^{-1}, S) \leq [b_{\mathcal{U}}(S^{-1}, S)]^2$.

Proof. By taking the closure of $\mathcal{U}$, we may assume without loss of generality that $\mathcal{U}$ is compact. Consider the operator $A = \int_{\mathcal{U}} U^*S^{-1*}S^{-1}U dU$ where dU is Haar measure on $\mathcal{U}$, normalized so that $\int_{\mathcal{U}} dU = 1$. Clearly $A \geq 0$. Further, using the invariance of dU, we have for any $V \in \mathcal{U}$,

$$V^{-1}AV = \int_{\mathcal{U}} V^*U^*S^{-1*}S^{-1}UV dU = \int_{\mathcal{U}} (UV)^*S^{-1*}S^{-1}(UV) d(UV) = A$$

so that any $V \in \mathcal{U}$ commutes with A and therefore $A \in \mathcal{D}$. Finally, for any $U \in \mathcal{U}$ and any vector $x \in \mathcal{H}$ we have $|S^{-1}USx|^2 \leq [b_{\mathcal{U}}(S^{-1}, S)]^2$ and $|x|^2 = |(S^{-1}U^*S)(S^{-1}USx)|^2 \leq [b_{\mathcal{U}}(S^{-1}, S)]^2 \, |S^{-1}USx|^2$, so we have the double inequality

$$(S^*ASx, x) = \int_{\mathcal{U}} (S^*U^*S^{-1*}S^{-1}USx, x) dU = \int_{\mathcal{U}} |S^{-1}USx|^2 dU$$

$$\lesseqgtr [b_{\mathcal{U}}(S^{-1}, S)]^{\pm 2} |x|^2 .$$

Setting $D = A^{\frac{1}{2}} \in \mathcal{D}$, we have that $\|DS\|^2 = \|S^*AS\| \leq [b_{\mathcal{U}}(S^{-1}, S)]^2$, $\|S^{-1}D^{-1}\|^2 = \|S^{-1}A^{-1}S^{-1*}\| = \|(S^*AS)^{-1}\| \leq [b_{\mathcal{U}}(S^{-1}, S)]^2$ and so $c_{\mathcal{D}}(S^{-1}, S) \leq [b_{\mathcal{U}}(S^{-1}, S)]^2$.

If we examine this proof, we see that all that was really needed was that the intersection of $\mathcal{D}$ with the closed (in the weak operator topology) convex hull of the operators U^*BU be non-void for every positive invertible B, for then any operator in the intersection may play the role of A: the bounds of (S^*ASx, x) are convex combinations of the bounds for $|S^{-1}USx|^2$. The von Neumann algebras $\mathcal{D}$ with this property for $\mathcal{U}$ the unitary group of $\mathcal{D}'$ include all $\mathcal{D}$ with only type I factors, as well as a few others. We refer to [2, pp. 168-171] for a discussion. We do not know of an example in which Theorem 7.1 fails, although we suspect that it fails to hold in complete generality. For the general case of commutative $\mathcal{U}$, see also [5, XV.6.1].

Since Theorem 7.1 does not depend on the nature of $\mathcal{U}$, one might wonder if some relation of the sort $c_{\mathcal{D}} \leq C b_{\mathcal{U}}$ held for arbitrary $\mathcal{U} \subset \mathcal{D}'$, say in the case of $\mathcal{D}$ maximal abelian. The answer is no.

7.2. EXAMPLE. Let $\mathcal{H}$ be of dimension 4^n, $\mathcal{D}$ the von Neumann algebra of diagonal matrices, and let $\mathcal{U}$ be the subgroup of diagonal matrices of the form $U = U_1 \otimes \ldots \otimes U_n$ where each U_ν is an arbitrary 4×4 diagonal unitary matrix. Then there is a number $\gamma > 0$ such that $\rho_{\mathcal{D},\mathcal{U}} \geq (4^n)^\gamma$. We need only consider the orthogonal, optimally conditioned, self-adjoint 4×4 projections P, Q constructed in [1, example 5.4] which have their diagonal entries in the proportion $2:1$; set $E = P \otimes \otimes \ldots \otimes P$, $F = Q \otimes \ldots \otimes Q$ (n factors) and note that the diagonal entries of E and F are in the proportion $2^n:1$ so that $c_{\mathcal{D}}(E, F) = \|E\|\,\|F\| = 1$. Since $\sup_\nu \|PVQ\| = b < 1$ as V runs over all 4×4 diagonal unitaries, we have for $U \in \mathcal{U}$ that $\|EUF\| = \|(P \otimes \ldots \otimes P)(U_1 \otimes \ldots \otimes U_n)(Q \otimes \ldots \otimes Q)\| = \prod_{\nu=1}^{n} \|PU_\nu Q\| \leq b^n$. We may choose $\gamma = -\log_4 b$. To see that $AU = UA$ for all $U \in \mathcal{U}$ implies $A \in \mathcal{D}$, it is probably easiest to write A as $\int_{\mathcal{U}} U^*AUdU$, coordinatize $\mathcal{U}$ in the obvious fashion as the product of n 4-tori, and calculate the integral directly to see that it is always diagonal.

By taking a direct sum of these examples, we have an example of $\mathcal{D}$ maximal abelian, $\mathcal{U}$ a determining unitary subgroup of $\mathcal{D}'$, such that $\rho_{\mathcal{D},\mathcal{U}} = \infty$.

In view of examples 6.4 and 7.2, the following simple theorem is not without interest, for it shows that $\rho_{\mathcal{D},\mathcal{U}}(R, S)$ can never be itself infinite.

7.2. THEOREM. If $b_{\mathcal{U}}(R, S) = 0$, then $c_{\mathcal{D}}(R, S) = 0$.

Proof. Let $\mathcal{H}_1$ be the closed subspace of $\mathcal{H}$ spanned by the ranges of $\{UR^*: U \in \mathcal{U}\}$, and let $\mathcal{H}_2$ be the closed subspace of $\mathcal{H}$ spanned by the ranges of $\{VS: V \in \mathcal{U}\}$. The hypothesis $RU^*VS = 0$ for all $U, V \in \mathcal{U}$ shows that $\mathcal{H}_1 \perp \mathcal{H}_2$. Let P, Q denote the orthogonal self-adjoint projections onto $\mathcal{H}_1, \mathcal{H}_2$. Since $\mathcal{H}_1$ and $\mathcal{H}_2$ are both invariant under every unitary in $\mathcal{U}$ and $\mathcal{U}$ is closed under adjoints, the subspaces $\mathcal{H}_1$ and $\mathcal{H}_2$ are reducing for every unitary in $\mathcal{U}$; it follows that P and Q commute with every $U \in \mathcal{U}$, and so $P, Q \in \mathcal{D}$. Let $D_\epsilon = \epsilon^{-1}p + \epsilon(I - P)$. Then $\|RD_\epsilon^{-1}\| = \|R(\epsilon P)\| \leq \epsilon\|R\|$ and $\|D_\epsilon S\| = \|\epsilon(I - P)S\| \leq \epsilon\|S\|$; as $\epsilon \downarrow 0$, $c_{\mathcal{D}}(R, S) = 0$.

We should like to note that it is not possible to generally get better estimates for $\rho_{\mathcal{D},\mathcal{U}}(S^{-1}, S)$ than $\rho_{\mathcal{D},\mathcal{U}}(R, S)$. Let $\hat{\mathcal{H}} = \mathcal{H} \oplus \mathcal{H}$, $\hat{\mathcal{D}} = \mathcal{D} \oplus \mathcal{D}$, $\hat{\mathcal{U}} = \mathcal{U} \oplus \mathcal{U}$. Suppose that $\rho_{\hat{\mathcal{D}},\hat{\mathcal{U}}}(S^{-1}, S) \leq C$ for all invertible operators S on $\hat{\mathcal{H}}$; the extremal case yields $\rho_{\hat{\mathcal{D}},\hat{\mathcal{U}}}(P, Q) \leq C$ for all pairs P, Q of orthogonal self-adjoint projections for which $c_{\hat{\mathcal{D}}}(P, Q) = 1$. Now let E, F be any pair of self-adjoint projections on $\mathcal{H}$ for which $c_{\mathcal{D}}(E, F) = 1$. Let $P = \frac{1}{2}\begin{pmatrix} E & E \\ E & E \end{pmatrix}$, $Q = \frac{1}{2}\begin{pmatrix} F & -F \\ -F & F \end{pmatrix}$. Clearly, P, Q are orthogonal, self adjoint projections on $\hat{\mathcal{H}}$. To see that $c_{\mathcal{D}}(P, Q) = 1$, we take $D = \begin{pmatrix} D_1 & 0 \\ 0 & D_2 \end{pmatrix}$ and we have

$$\|PD^{-1}\|\,\|DQ\| = \frac{1}{\sqrt{2}}(\|ED_1^{-1}\|^2 + \|ED_2^{-1}\|^2)^{\frac{1}{2}} \cdot \frac{1}{\sqrt{2}}(\|D_1F\|^2 + \|D_2F\|^2)^{\frac{1}{2}}$$

$$\geq \|ED_1^{-1}\|\,\|ED_2^{-1}\|\,\|D_1F\|\,\|D_2F\| \geq 1.$$

To estimate $b_{\mathcal{U}}(E, F)$ in terms of $b_{\hat{\mathcal{U}}}(P, Q) \geq C^{-1}$, we take $\hat{U} = \begin{pmatrix} U_1 & 0 \\ 0 & U_2 \end{pmatrix}$ so that

$$\|PUQ\| = \left\|\frac{1}{4}\begin{pmatrix} E(U_1 - U_2)F & E(U_2 - U_1)F \\ E(U_2 - U_1)F & E(U_1 - U_2)F \end{pmatrix}\right\| = \|E(U_1 - U_2)F\| \leq b_{\mathcal{U}}(E, F).$$

It follows that $b_{\mathcal{U}}(E, F) \geqslant C^{-1}$ and thus we see that $\rho_{\mathcal{D},\mathcal{U}} \leqslant \sup \rho_{\hat{\mathcal{D}},\hat{\mathcal{U}}}(S^{-1}, S)$.

The outstanding problem remaining seems to be the following: given $\mathcal{H}$ finite dimensional and $\mathcal{D}$ presented as at the beginning of §6, to give sharp estimates of $\rho_{\mathcal{D},\mathcal{U}}$ in terms of the numbers k, m_k, and n_k.

Added in proof. I have discovered that Lemma 3.2 is a special case of Theorem 3.4 (2) of J. Bendat and S. Sherman, Monotone and convex operator functions, Trans. Amer. Math. Soc., 79 (1955), pp. 58-71: $\phi(x) = 1/x$ is operator-convex on any subinterval of $(0, \infty)$.

REFERENCES

1. C. McCarthy and G. Strang, Optimal conditioning of matrices, to appear.
2. J. T. Schwartz, W*-Algebras, Gordon and Breach, 1967.
3. C. McCarthy, c_p, Israel J. Math., 5 (1967), pp. 249-271.
4. G. E. Forsythe and E. G. Straus, On best conditioned matrices, Proc. Amer. Math. Soc., 6 (1955), pp. 340-345.
5. N. Dunford and J. T. Schwartz, Linear Operators III, to appear.

ON THE SMOOTHNESS OF FUNCTIONS SATISFYING CERTAIN INTEGRAL INEQUALITIES

Adriano M. Garsia – *University of California, San Diego*

CONTENTS

This will be mostly an expository presentation of the results of an investigation organized by the present author concerning the smoothness of functions $f(x)$ on $[0, 1]$ which satisfy inequalities of the type

$$I_{\Psi, p}(f) = \int_0^1 \int_0^1 \Psi((f(x) - f(y))/p(x - y))dxdy \leqslant B < \infty.$$

where $\Psi(u)$ is defined on $(-\infty, +\infty)$ and

(a) $\Psi(u) = \Psi(-u) \geqslant 0,$

(b) $\Psi(u) \uparrow \infty$ as $u \uparrow \infty,$

while $p(u)$ is defined and continuous on $[-1, 1]$ and

(a) $p(u) = p(-u) \geqslant 0,$

(b) $p(u) \downarrow 0$ as $u \downarrow 0.$

The research we shall describe here has been carried out during a span of two years and involves contributions by the author, C. Preston, E. Rodemich, H. Rumsey, Jr. and H. Taylor.

In our presentation, we shall try as much as possible to adhere very closely to the original lines of thought which led to these results. Furthermore, to simplify our exposition and in order not to lose ourselves in technicalities, we shall occasionally have to refer the reader to the original manuscripts for the detailed proofs of some of our assertions.

§1. THE POINT OF DEPARTURE. The original stimulus for our work came from a study of the convergence behavior of the Karhunen-Loève expansion

$$X_t(\omega) = \sum_{\nu=1}^{\infty} \sqrt{\lambda_\nu}\varphi_\nu(t)\theta_\nu(\omega) \tag{1.1}$$

of a Gaussian process on $[0, 1]$.

To introduce our notation, and for the sake of completeness, it is well to review here some of the basic facts concerning such processes.

Here and in the following $X_t(\omega)$ will denote a mean continuous, zero-mean, measurable and separable real Gaussian process on $[0, 1]$. It is well-known that such a process has a covariance function

$$R(s, t) = E(X_s X_t) \tag{1.2}$$

which is continuous and positive definite on the square $S = [0, 1] \times [0, 1]$.

We also set for any $(s, t) \in S$

$$\Delta R(s, t) = R(s, s) + R(t, t) - 2R(s, t) = E(|X_t - X_s|^2) \tag{1.3}$$

By definition, $\{X_t\}$ is mean continuous if and only if

$$\lim_{|t-s|\to 0} E(|X_t - X_s|^2) = 0,$$

and this condition is easily seen to imply that

$$\text{(1.4)} \quad \begin{array}{ll} \text{(a)} & \sup\{R(s, s): s \in [0, 1]\} < \infty, \\ \text{(b)} & \lim_{|s-t|\to 0} \Delta R(s, t) = 0. \end{array}$$

From 1.3 a simple manipulation and Schwarz's inequality yield

$$|R(s_1, t_1) - R(s_2, t_2)| \leq \sqrt{R(s_1, s_1)}\sqrt{\Delta R(t_1, t_2)} + \sqrt{R(t_2, t_2)}\sqrt{\Delta R(s_1, s_2)}$$

Thus, the continuity of $R(s, t)$ on S follows from (1.4) (a) and (b).

On the other hand, the identity

$$\int_0^1 \int_0^1 R(s, t)\varphi(s)\overline{\varphi(t)}dsdt = E(|\int_0^1 \varphi(s)X_s(\omega)ds|^2), \tag{1.5}$$

which is valid for any function $\varphi(t)$ that is square integrable on $[0, 1]$, shows that $R(s, t)$ is a positive definite kernel.

This given, $R(s, t)$ must admit a Mercer's expansion

$$R(s, t) = \sum_{\nu=1}^{\infty} \lambda_\nu \varphi_\nu(s)\varphi_\nu(t) \tag{1.6}$$

which converges uniformly in the square S, where $\{\varphi_n(s)\}$ is an orthonormal set of eigenfunctions of the integral equation

$$\lambda\varphi(s) = \int_0^1 R(s, t)\varphi(t)dt.$$

From (1.6) we easily derive the relations

$$\sum_{\nu=1}^{\infty} \lambda_\nu = \int_0^1 R(s, s)ds, \tag{1.7}$$

$$\Delta R(s, t) = \sum_{\nu=1}^{\infty} \lambda_\nu(\varphi_\nu(s) - \varphi_\nu(t))^2. \tag{1.8}$$

Both series here are absolutely convergent since the λ_ν's are positive numbers.

Finally, any function $\varphi \in \mathcal{L}_2(0, 1)$ which is orthogonal to all φ_n's must satisfy (in view of (1.5) and (1.6)) the relation

$$E(|\int_0^1 X_s \varphi(s)ds|^2) = 0,$$

and from this fact it can be derived that with probability one the paths $X_t(\omega)$ of the process admit an $\mathcal{L}_2$ expansion of the form

$$X_t(\omega) \sim \sum_{\nu=1}^{\infty} \varphi_\nu(t) \int_0^1 \varphi_\nu(s)X_s(\omega)dt.$$

Furthermore, the coefficients

$$\int_0^1 \varphi_\nu(s)X_s(\omega)ds,$$

by the orthogonality of the φ_n's, constitute an orthogonal set of random variables. If we set

$$\theta_\nu(\omega) = \frac{1}{\sqrt{\lambda_\nu}} \int_0^1 \varphi_\nu(s)X_s(\omega)ds,$$

the system $\{\theta_\nu(\omega)\}$ will also be orthonormal. Thus, in the case that the process X_t is Gaussian, the θ_ν's turn out to be a sequence of independent, standard normal variables.

From the classical results of Kolmogorov, we can then derive that the series

$$\sum_{\nu=1}^{\infty} \sqrt{\lambda_\nu}\,\theta_\nu(\omega)\varphi_\nu(t) \tag{1.9}$$

will converge, for every $t \in [0, 1]$, to $X_t(\omega)$ with probability one, and thus, by Fubini's theorem, it will converge with probability one a.e. in t.

Given such a state of affairs, it is tempting to use the expansion (1.9) for the construction of a model of the process, given its convariance function. However, these circumstances are not usually exploited for existence purposes, even in the Gaussian case.

The main difficulty consists in the fact that, excepted the trivial case when there is only a finite number of nonzero λ_ν's, very little has been known about the general behavior of the partial sums

$$X_t^{(n)}(\omega) = \sum_{\nu=1}^{n} \lambda_\nu \varphi_\nu(t)\theta_\nu(\omega)$$

beyond their a.e. convergence with probability one. Consequently, there appears to be no natural way to select a limit $X_t(\omega)$ (for all t) which will give us the desired *measurable* and *separable* model of the process.

Nevertheless, this is precisely the path chosen by N. Wiener in his famed construction of a mathematical model for Brownian motion.

Over the past few years various smoothness conditions on $R(s, t)$ have been shown to imply the existence of a model of the process which has continuous paths with probability one. (See, for instance [3] or [5].)

In other words, in some cases, the series in (1.9), with probability one admits a "limit" which is a continuous function of t. It would be quite remarkable if this *continuity* were not a consequence of the *uniform* convergence of the series itself! However, things are not as bad as all that. It can indeed be shown (for a proof see [7]) that the partial sums

$$X_t^{(n)}(\omega) = \sum_{\nu=1}^{n} \sqrt{\lambda_\nu}\,\varphi_\nu(t)\theta_\nu(\omega) \tag{1.10}$$

do converge uniformly in t with probability one if and only if there is a measurable and separable model of the process whose paths are continuous with probability one.

This result is certainly necessary if there is to be any hope that, under rather general circumstances, the series in (1.9) may actually be used for the construction of such models.

It is thus clear that it should make a very interesting program to try and see in what way smoothness conditions on $R(s, t)$ do directly affect the behavior of the partial sums in (1.10). This was essentially the original motivation for our investigation.

§2. WIENER'S APPROACH. The inspiration in our investigation came from N. Wiener's work ([11], Chapter IX) on Brownian motion, and it is good to give here a sketch of Wiener's line of reasoning.

Note first that for the Brownian motion process we have

$$R(s, t) = \min(s, t), \tag{2.1}$$

$$\varphi_\nu(t) = \sqrt{2}\sin(\nu + \tfrac{1}{2})\pi t, \tag{2.2}$$

and

$$\lambda_\nu = 2/(2\nu + 1)^2\pi^2. \tag{2.3}$$

This given, Wiener works with the subsequence

$$X_t^{(n_k)}(\omega) = \sum_{\nu=1}^{n_k} \sqrt{\lambda_\nu}\varphi_\nu(t)\theta_\nu(\omega),$$

for $n_k = 2^k$. His program is to show that if we set

$$M_k(\omega) = \max\{|\sum_{\nu=n_k+1}^{n_{k+1}} \sqrt{\lambda_\nu}\varphi_\nu(t)\theta_\nu(\omega)| : 0 \leq t \leq 1\},$$

we necessarily have

$$\sum_{k=1}^{\infty} E(M_k) < \infty. \tag{2.4}$$

This result, of course, only implies that the *subsequence* $X_t^{(m_k)}$ converges uniformly, with probability one, but that was sufficient for Wiener's purposes.

We should point out, however, that only a very small addition to Wiener's argument yields the stronger result

$$\sum_{k=1}^{\infty} E(M_k^*) < \infty \tag{2.5}$$

where

$$M_k^*(\omega) = \max\{\max\{|\sum_{\nu=n_k+1}^{n} \sqrt{\lambda_\nu}\varphi_\nu(t)\theta_\nu(\omega)| : n_k < n \leq n_{k+1}\} : 0 \leq t \leq 1\}.$$

And 2.5 does imply the a.s. uniform convergence of the unrestricted sequence $X_t^{(n)}$.

We can obtain (2.4) by estimating each term $E(M_k)$ separately. To this end, for given $n < m$ set

$$P_t(\omega) = \sum_{n<\nu\leq m} \sqrt{\lambda_\nu}\varphi_\nu(t)\theta_\nu(\omega).$$

Then observe that since $\{\theta_n\}$ is a system of independent standard Gaussian variables, we must have

$$E(\exp(1/4)[p_t^2 / \sum_{n<\nu\leq m} \lambda_\nu\varphi_\nu^2(t)]) \leq \sqrt{2}. \; * \tag{2.6}$$

* This is a simple consequence of the identity $\int_{-\infty}^{+\infty} \exp \tfrac{1}{4}\{x^2\} e^{-x^2/2} \frac{dx}{\sqrt{2\pi}} = \sqrt{2}$.

If we integrate this relation with respect to t from 0 to 1 and use (2.2) we get

$$\text{(2.7)} \qquad E\left(\int_0^1 \exp(1/8)[p_t^2/\sum_{n<\nu\leq m} \lambda_\nu]\, dt\right) \leq \sqrt{2}.$$

What we do need here is an estimate on $E(M)$ where

$$\text{(2.8)} \qquad M(\omega) = \max_{0\leq t\leq 1} |P_t(\omega)|,$$

however, (2.7) is not too far from that. Indeed, we might regard (2.7) as giving an estimate on the expected value of a "high order" norm of $P_t(\omega)$ (as a function of t) and there is a rather clever way to deduce from it an estimate on the $\mathcal{L}_\infty$ norm of P_t.

The idea is that if P_t does not vary too rapidly in t, for a given ω, it will remain close to its maximum in a small t interval, and thus the $\mathcal{L}_\infty$ norm of P_t cannot be too large without forcing also the integral.

$$\int_0^1 \exp(1/8)[p_t^2/\sum_{n<\nu\leq m} \lambda_\nu]\, dt$$

to be large as well.

The crucial step is to use Bernstein's inequality

$$\text{(2.9)} \qquad \|P'\|_\infty \leq \text{degree } P\ \|P\|_\infty,$$

which is valid for every trigonometric polynomial P, as a ***control on the variability*** of P_t.

A simple use of (2.9) yields immediately the inequality

$$(1/2m)\exp(1/32)[\|P\|_\infty^2/\sum_{n<\nu\leq m} \lambda_\nu] \leq \int_0^1 \exp(1/8)[p_t^2/\sum_{n<\nu\leq m} \lambda_\nu]\, dt.$$

This given, taking (2.3) into account, the result in (2.4) can be easily derived.

§3. THE FIRST BASIC LEMMA. In trying to adapt the above argument to the general case we are led to find the analogue of Bernstein's inequality for linear combinations of eigenfunctions ("polynomials") corresponding to a given kernel. Or, at any rate, that failing, we should try to find some other measure of the "variability" of these polynomials in terms of a measure of the variability of $R(s, t)$.

A most natural measure of the variability of R is the function

$$\Delta R(s,t) = \sum_{\nu=1}^{\infty} \lambda_\nu (\varphi_\nu(s) - \varphi_\nu(t))^2$$

or, better yet, the function

$$p(u) = \max_{|s-t|\leq u} \sqrt{\Delta R(s,t)}.$$

A view of the inequality in (2.6) might suggest observing that we do have as well

$$E(\exp(1/4)[(P_s - P_t)^2 / \sum_{n<\nu\leq m} \lambda_\nu(\varphi_\nu(s) - \varphi_\nu(t))^2]) \leq \sqrt{2}.$$

Thus, *a fortiori*

$$E(\exp(1/4)[(P_s - P_t)/p(s - t)]^2) \leq \sqrt{2}.$$

Integrating this inequality over $S = [0, 1] \times [0, 1]$ we finally get

$$E(\int_0^1 \int_0^1 \exp(1/4)[(P_s - P_t)/p(s - t)]^2 dsdt) \leq \sqrt{2}.$$

A very natural question, in view of the arguments in the previous section, then arises. Suppose we know that a certain continuous function $f(t)$ on $[0, 1]$ satisfies two inequalities of the type

$$(3.1) \qquad \int_0^1 \exp(1/c)(f(t)/A)^2 dt \leq c$$

$$(3.2) \qquad \int_0^1 \int_0^1 \exp(1/c)((f(s) - f(t))/p(s - t))^2 dsdt \leq B$$

for some universal constant c. For what functions $p(u)$ can we deduce, purely from (3.1) and (3.2) an *a priori* estimate for the $\mathcal{L}_\infty$ norm of f in terms of A and B?

This question led to the basic lemma we shall present in this section.

To make our story short, we may observe that, in the presence of (3.1), the estimates we desire concerning $f(t)$ can be immediately derived, if we could only obtain from (3.2) an *a priori* estimate for the oscillation of f on $[0, 1]$, namely the quantity

$$(3.3) \qquad |\max_{0\leq t\leq 1} f(t) - \min_{0\leq t\leq 1} f(t)|,$$

in terms of B.

At first analysis, such a result appears doubtful, for by a change of scale in the argument, such an *a priori* estimate could be used to obtain an inequality for the oscillation of f in any subinterval $[x, y]$ in terms of the corresponding integral

$$\int_x^y \int_x^y \exp(1/c)[(f(s) - f(t))/p(s - t))]^2 dsdt.$$

This, in turn should imply that only *essentially* continuous functions $f(t)$ could satisfy such an inequality as (3.2). And the latter would be quite remarkable.

However, for some functions $p(u)$, this is exactly what happens. In fact, a rather general result to this effect does indeed hold true.

To be precise, let here and in the following $\Psi(u)$ and $p(u)$ be two functions, defined on $(-\infty, +\infty)$ and $[-1, 1]$ respectively, such that

(3.4)

(a) $\Psi(u) = \Psi(-u) \geq 0$ for every $u \in (-\infty, +\infty)$,

(b) $\Psi(u) \uparrow \infty$ as $u \uparrow \infty$,

and

$$(3.5)\qquad \begin{array}{ll}(a) & p(u) = p(-u) \geqslant 0 \text{ for every } u \in [-1, 1],\\ (b) & p(u) \downarrow 0 \text{ as } u \downarrow 0.\end{array}$$

This given, let us set for each $u \geqslant \Psi(0)$

$$\Psi^{-1}(u) = \sup\{v: \Psi(v) \leqslant u\},$$

and for $0 \leqslant u \leqslant p(1)$

$$p^{-1}(u) = \max\{v: p(v) \leqslant u\}.$$

The following result holds.

LEMMA 3.1. Let $f(t)$ be defined and continuous on $[0, 1]$ and suppose that

$$(3.6)\qquad \int_0^1 \int_0^1 \Psi((f(s) - f(t))/p(s - t))dsdt \leqslant B < \infty;$$

then for all $s, t \in [0, 1]$ we have

$$(3.7)\qquad |f(s) - f(t)| \leqslant 8 \int_0^{|s-t|} \Psi^{-1}(4B/u^2)dp(u).$$

A very short proof of this result can be found in [7]. Here we shall eventually deduce another proof via a longer but perhaps more illuminating path.

First of all, observe that to obtain Lemma 3.1, we need only show (3.7) for $s = 1$ and $t = 0$. I.e. it is sufficient to prove

$$(3.8)\qquad |f(1) - f(0)| \leqslant 8 \int_0^1 \Psi^{-1}(4B/u^2)dp(u),$$

for any f satisfying (3.6).

Indeed, applying (3.8) to the function

$$\tilde{f}(\tilde{t}) = f(s + \tilde{t}(s - t)) \quad \tilde{t} \in [0, 1]$$

with Ψ unchanged but p replaced by

$$\tilde{p}(u) = p(u|s - t|)$$

and B replaced by

$$\tilde{B} = \int_0^1 \int_0^1 \Psi((\tilde{f}(\tilde{s}) - \tilde{f}(\tilde{t}))/\tilde{p}(\tilde{s} - \tilde{t}))d\tilde{s}d\tilde{t} \leqslant B/|s - t|^2,$$

we deduce

$$|f(s) - f(t)| = |\tilde{f}(1) - \tilde{f}(0)| \leqslant 8 \int_0^1 \Psi^{-1}(4B/u^2|s - t|^2)dp(u|s - t|).$$

However, modulo a change of variables, this is precisely the inequality in (3.7).

§4. THE CASE OF MONOTONE FUNCTIONS. During the first attempt at proving Lemma 3.1, E. Rodemich discovered that (3.8) admits a rather simple proof if f, in addition to satisfying (3.6), is also assumed to be monotone in [0, 1]. Indeed, in this case a sharper inequality holds true, namely

LEMMA 4.1. Let f(t) be nonincreasing in [0, 1] and let

$$\int_0^1 \int_0^1 \Psi((f(s) - f(t))/p(s - t))dsdt \leq B < \infty; \tag{4.1}$$

then

$$f(0^+) - f(1^-) \leq 8 \int_0^{1/2} \Psi^{-1}(B/u^2)dp(u). \tag{4.2}$$

Proof. We introduce a sequence of numbers $\{t_n\}$ decreasing to zero by setting

(4.3)
(a) $$t_0 = 1/2,$$
(b) $$t_{n+1} = p^{-1}(p(t_n)/2), \text{ for every } n \geq 0.$$

For each $n \geq 0$, we let Δ_n be the triangle

$$\Delta_n = \{(s, t): t_n \leq s \leq t_n + t_{n+1},\ s - t_n \leq t \leq t_{n+1}\}.$$

Clearly Δ_n is contained in the triangle $\{(s, t): 0 \leq t \leq s\}$ and thus

$$\iint_{\Delta_n} \Psi((f(s) - f(t))/p(s - t))dsdt \leq B/2. \tag{4.4}$$

On the other hand in Δ_n we have

$$t \leq t_{n+1} < t_n \leq s$$

and

$$t_n - t_{n+1} \leq s - t \leq t_n.$$

Thus, by the monotonicity of Ψ, p and f, we get

$$\iint_{\Delta_n} \Psi((f(s) - f(t))/p(s - t))dsdt \geq (1/2)t_{n+1}^2 \Psi((f(t_{n+1}) - f(t_n))/p(t_n)). \tag{4.5}$$

Combining (4.5) and (4.4) we easily obtain *

$$f(t_{n+1}) - f(t_n) \leqslant \Psi^{-1}(B/t_{n+1}^2)p(t_n). \tag{4.6}$$

Since from (4.3) (b), $p(t_n) = 4[p(t_{n+1}) - p(t_{n+2})]$, we can replace the inequality in (4.6) by

$$f(t_{n+1}) - f(t_n) \leqslant 4 \int_{t_{n+2}}^{t_{n+1}} \Psi^{-1}(B/u^2)dp(u). \tag{4.7}$$

Upon summing for $n \geqslant 0$, and passing to the limit, we get

$$f(0^+) - f(1/2) \leqslant 4 \int_0^{1/2} \Psi^{-1}(B/u^2)dp(u). \tag{4.8}$$

A use of this inequality with $f(t)$ replaced by $-f(1 - t)$ yields

$$f(1/2) - f(1^-) \leqslant 4 \int_0^{1/2} \Psi^{-1}(B/u^2)dp(u), \tag{4.9}$$

and (4.2) is obtained upon summing (4.8) and (4.9).

REMARK 4.1. A corollary of the above lemma is the result that a monotone $f(t)$ satisfying (3.6) is necessarily continuous within $(0, 1)$ when

$$\int_0^\delta \Psi^{-1}(B/u^2)dp(u) < \infty.$$

Indeed, by the change of variable argument used in the previous section, we can show that the inequality

$$|f(x) - f(y)| \leqslant 8 \int_0^{|x-y|} \Psi^{-1}(B/u^2)dp(u)$$

must hold whenever x and y are points of continuity of f.

§5. THE CONJECTURE. We have thus seen that Lemma 3.1 is true at least in the case that $f(x)$ is monotone. However, this fact suggests a very elegant approach to Lemma 3.1.

Let us recall that, given a measurable function $f(x)$ on $[0, 1]$ we can associate to it a monotone nonincreasing function $f^*(x)$ on $[0, 1]$ which is uniquely characterized by the condition that

$$m\{x: f(x) > \lambda\} = m\{x: f^*(x) > \lambda\} \text{ for every } \lambda. \tag{5.1}$$

* Provided B is sufficiently large.

The function $f^*(x)$ is usually referred to as the nonincreasing "rearrangement" of $f(x)$, and it is a classical tool (see [16], V.1, §13) for it carries important information concerning $f(x)$ itself.

The two basic properties of $f^*(x)$, as related to $f(x)$, which we wish to recall here are the following

$$(5.1) \qquad \begin{aligned} &(a) \qquad f^*(0^+) = \operatorname*{ess\,sup}_{x\in[0,1]} f(x) \\ &(b) \qquad f^*(1^-) = \operatorname*{ess\,inf}_{x\in[0,1]} f(x) \end{aligned}$$

and

$$(5.2) \qquad \int_0^1 \varphi(f^*)dx = \int_0^1 \varphi(f)dx,$$

the latter holding true for every function $\varphi(u)$ which is locally of bounded variation on $(-\infty, +\infty)$.

This given, it is tempting to make the following:

Conjecture. When f is measurable and Ψ and p satisfy the conditions in (3.4) and (3.5) we have

$$(5.3) \qquad \int_0^1 \int_0^1 \Psi((F^*(x) - f^*(y))/p(x-y))dxdy \leqslant$$

$$\leqslant \int_0^1 \int_0^1 \Psi((f(x) - f(y)/p(x-y))dxdy.$$

Indeed, if this conjecture is true, Lemma 3.1 follows immediately from Lemma 4.1. For, by (3.6), (5.3) and (4.2) we immediately derive

$$(5.4) \qquad \operatorname*{ess\,sup}_{x\in[0,1]} f(x) - \operatorname*{ess\,inf}_{x\in[0,1]} f(x) \leqslant 9\int_0^{1/2} \Psi^{-1}(B/u^2)dp(u),$$

and thus, *a fortiori*, if $f(x)$ is continuous in $[0, 1]$ we get

$$|f(1) - f(0)| \leqslant 8 \int_0^{1/2} \Psi^{-1}(B/u^2)dp(u).$$

This relation, as we have seen, does imply (3.7). Thus Lemma 3.1 would be established in full generality.

Unfortunately, to this date we have been unable to determine whether or not this conjecture is true in the generality required by the argument above, so that in [7] an entirely different proof of Lemma 3.1 had to be derived.

The conjecture would be a rather interesting result for its own merits, for as we shall see it has several important consequences which go beyond the scope of Lemma 3.1 itself. However, before dwelling any further on these matters it is good to show the next information which Lemma 3.1 yields about the behavior of the partial sums of the Karhunen-Loève expansion of Gaussian processes.

REMARK 5.1. Before closing this section we should point out that from (5.4) the following sharpening of Lemma 3.1 can be easily obtained.

LEMMA 5.1. Let $f(x)$ be a measurable function On $[0, 1]$ and suppose that the function

$$\int_0^1 \int_0^1 \Psi((f(s) - f(t))/p(s - t))dsdt \leq B < \infty;$$

then if

$$\int_0^{1/2} \Psi^{-1}(B/u^2)dp(u) < \infty$$

the function $f(t)$ is essentially continuous. More precisely there is a function $\tilde{f}(t)$ equal to $f(t)$ almost everywhere in $[0, 1]$ which satisfies the inequality.

$$|\tilde{f}(x) - \tilde{f}(y)| \leq 8 \int_0^{|x-y|} \Psi^{-1}(B/u^2)dp(u)$$

for every $(s, t) \in [0, 1]$.

§6. APPLICATIONS TO GAUSSIAN PROCESSES. Lemma 3.1 yields uniform convergence results for the Karhunen-Loève expansion of Gaussian processes, which are as good as can be obtained (see [7]), in a swift and elegant fashion.

To see this let $R(s, t)$ be a continuous and positive definite kernel on the square $S = [0, 1] \times [0, 1]$ and let

$$R(s, t) = \sum_{\nu=1}^{\infty} \lambda_\nu \varphi_\nu(s)\varphi_\nu(t)$$

be its Mercer's expansion on S. Set

$$\Delta R(s, t) = \sum_{\nu=1}^{\infty} \lambda_\nu(\varphi_\nu(s) - \varphi_\nu(t))^2$$

and

$$p(u) = \max\{\sqrt{\Delta R(s, t)}: |s - t| \leq u\}.$$

Furthermore, let $\{\theta_n(\omega)\}$ be a sequence of independent, standard Gaussian random variables. The following result holds:

THEOREM 6.1. Suppose that

(6.1) $$\int_0^1 \sqrt{\log 1/u}\, dp(u) < \infty$$

Then, with probability one, the partial sums

(6.2) $$X_t^{(n)}(\omega) = \sum_{\nu=1}^{n} \sqrt{\lambda_\nu}\varphi_\nu(t)\theta_\nu(\omega)$$

converge uniformly in $[0, 1]$. Furthermore, there is a random variable $B(\omega)$ with finite expectation such that, for all n and for all $s, t \in [0, 1]$ we have

$$(6.3) \qquad |X_s^{(n)}(\omega) - X_t^{(n)}(\omega)| \leq 16 \int_0^{|s-t|} \sqrt{\log 4B(\omega)/u^2}\, dp(u).$$

Proof. Let us observe that we need only show that the random variable

$$B(\omega) = \sup \int_0^1 \int_0^1 \exp\{(X_s^{(n)}(\omega) - X_t^{(n)}(\omega))/2p(s-t)\}^2 dsdt$$

has finite expectation, for an application of Lemma 3.1 immediately gives (6.3). This given, at any ω where

$$(6.4) \quad \begin{array}{ll} \text{(a)} & B(\omega) < \infty, \\ \text{(b)} & \sum_{\nu=1}^{\infty} \lambda_\nu [\theta_\nu(\omega)]^2 < \infty, \end{array}$$

the partial sums in (6.2) will be, as functions of t, both equicontinuous and convergent in the $\mathcal{L}_2$ sense. Thus for such an ω, the sequence $\{X_t^{(n)}(\omega)\}$ converges uniformly. On the other hand, if $B(\omega)$ has finite expectation, (6.4), (a) will necessarily hold with probability one, while (6.4), (b) trivially holds with probability one as well since

$$E(\sum_{\nu=1}^{\infty} \lambda_\nu \theta_\nu^2(\omega)) = \sum_{\nu=1}^{\infty} \lambda_\nu < \infty.$$

Let us now show that $B(\omega)$ has finite expectation. To this end, for fixed s and t set

$$P_n(\omega) = \exp \tfrac{1}{2}\{(X_s^{(n)}(\omega) - X_t^{(n)}(\omega))/2p(s-t)\}^2.$$

By the convexity of e^{u^2} and the definition of $X_s^{(n)}(\omega), X_t^{(n)}(\omega)$, it is easily shown that the sequence $\{P_n\}$ is a submartingale. From the classical martingale inequalities (see [4]) we can then deduce that

$$(6.5) \qquad E(\max\{P_m^2(\omega) : m \leq n\}) \leq 4E(P_n^2(\omega)).$$

On the other hand since for each n the random variable

$$(X_s^{(n)}(\omega) - X_t^{(n)}(\omega))/p(s-t)$$

is Gaussian, has mean zero and variance less than or equal to one, we must have

$$(6.6) \qquad E(p_n^2(\omega)) \leq \sqrt{2}.$$

Combining (6.5) with (6.6) and integrating with respect to s and t from zero to one, by Fubini's theorem we obtain

$$E(\int_0^1 \int_0^1 \max\{\exp \tfrac{1}{4}\{(X_s^{(m)}(\omega) - X_t^{(m)}(\omega))/p(s-t))\}^2 : m \leq n\} dsdt) \leq 4\sqrt{2}.$$

Upon letting $n \uparrow \infty$ in this relation and using the monotone convergence theorem, we easily derive the inequality

$$E(B(\omega)) \leq 4\sqrt{2}.$$

This completes the proof of the theorem.

Remarks. These methods yield results of similar nature concerning mean continuous processes that are not necessarily Gaussian. However, for such processes the conditions on $R(s, t)$ which assure the a.s. uniform convergence of the Karhunen-Loève expansion are necessarily more stringent. The reader is referred to [7] for some further applications of Lemma 3.1.

Before closing this section we wish to write our estimate for the modulus of continuity of sample paths in a slightly different form. To this end, we pass to the limit in (6.3) as $n \to \infty$, and use the inequality $\sqrt{a+b} \leq \sqrt{a} + \sqrt{b}$ * to obtain

$$(6.7) \qquad |X_s(\omega) - X_t(\omega)| \leq 16\sqrt{\log 4B(\omega)}\, p(s-t) + \\ + 16 \int_0^{|s-t|} \sqrt{\log 1/u^2}\, dp(u).$$

Note now that the significant term in this inequality is purely deterministic. Thus in general the influence of the particular path may make itself felt here only for relatively large values of $|s-t|$.

For the Brownian motion process (6.7) becomes

$$|X_s(\omega) - X_t(\omega)| \leq 16\sqrt{|s-t| \log 4B(\omega)} + 16\sqrt{2} \int_0^{|s-t|} \sqrt{(1/u) \log 1/u}\, du.$$

This inequality is rather interesting in view of the remarks at the end of Chapter IX in [11].

§7. LOOKING BEYOND THE FIRST BASIC LEMMA. CARLESON'S INEQUALITY.

The conclusion of Lemma 3.1 is vacuous if

$$(7.1) \qquad \int_0^1 \Psi^{-1}(4B/u^2)dp(u) = \infty.$$

Yet it seems hard to believe that when this happens nothing at all may be inferred about a function which satisfies a condition such as

$$(7.2) \qquad \int_0^1 \int_0^1 \Psi((f(x) - f(y))/p(x-y))dxdy \leq B < \infty.$$

However, it is not difficult to show that, if (7.1) holds, the condition (7.2) can be

* This loses a factor of $\sqrt{2}$ at the most.

satisfied by discontinuous functions. Thus, in the presence of (7.1), whatever conclusions may be derived from (7.2) regarding $f(x)$, they must be of a rather different nature than (3.7).

These considerations led us to seek for a more general result concerning functions satisfying (7.2) which would reduce to Lemma 3.1 when

$$\int_0^1 \Psi^{-1}(4B/u^2)dp(u) < \infty.$$

We believe we have found such a result, but before presenting it we shall describe the line of reasoning which led us to its discovery.

Our inspiration here came from Carleson's works [1] and [2] on capacities and exceptional sets. Carleson, among other things, is concerned with functions on $(-\infty, +\infty)$, periodic of period 2π, satisfying conditions of the type

$$\text{(7.3)} \qquad \int_{-\pi}^{\pi} f^2(x)dx + \int_{-\pi}^{\pi}\int_{-\pi}^{\pi} |(f(x) - f(y))/p(x-y)|^2 dxdy < \infty.$$

Such conditions can also be written in the form

$$\text{(7.4)} \qquad \Sigma\, |c_n|^2 \lambda_n < \infty,$$

where

$$c_n = \frac{1}{2\pi}\int_{-\pi}^{\pi} e^{-inx} f(x)dx.$$

The sequences $\{\lambda_n\}$ allowed in (7.4) are usually taken to be such that the kernel

$$\text{(7.5)} \qquad k(x) \sim \sum_n e^{inx}/\lambda_n$$

has several desirable properties including its being nonnegative, integrable, symmetric and increasing to $+\infty$ as $|x| \downarrow 0$ in $[-\pi, \pi]$.

It is easily seen that the class of functions satisfying (7.4), for a fixed $\{\lambda_n\}$, forms a Hilbert subspace of $\mathcal{L}_2(-\pi, \pi)$ under the norm

$$\text{(7.6)} \qquad \sqrt{\sum_n |c_n|^2 \lambda_n}.$$

A general theory of such spaces was initiated by Frostman [6]. The classical example comes from Dirichlet's condition

$$\text{(7.7)} \qquad \int_{-\pi}^{\pi}\int_{-\pi}^{\pi} ((f(x) - f(y))/(x-y))^2 dxdy < \infty$$

which is equivalent to

$$\text{(7.8)} \qquad \sum_n |c_n|^2 n < \infty.$$

Indeed, the functions satisfying (7.7) are precisely those functions on the unit circle whose harmonic continuation in the unit disk has finite Dirichlet integral. Perhaps, for this reason, these spaces are referred to as "Dirichlet spaces".

It is good to give here a brief review of the basic notions of the theory. For further information and details the reader may consult Carleson's monograph [1] or better yet Preston's thesis [12].

First of all, if K and $\{\lambda_n\}$ are related by (7.5), the Hilbert space corresponding to (7.4) will be referred to as $\mathcal{H}_K$ and the norm (7.6) of an $f \in \mathcal{H}_K$ will be denoted by $\|f\|_K$.

Further, a function $u_\mu(x)$, periodic of period 2π of the form

$$u_\mu(x) = \int_{-\pi}^{\pi} K(x-t)d\mu(t)$$

where μ is a nonnegative mass distribution on $[-\pi, \pi]$ is called a "potential". For a given Borel set $E \subset [-\pi, \pi]$ one defines the "capacity" of E with respect to K and denotes it by $C_K(E)$:

$$C_K(E) = \sup\{\mu(E): u_\mu \leq 1\}, \tag{7.9}$$

where the supremum is taken over all μ whose support is contained in E. This is all done in analogy with the definitions of classical potential theory. A great deal of the efforts in the theory of Dirichlet spaces has been in the direction of making these spaces as general as possible, by weakening the definitions as far as is permissible without losing the basic properties of the classical notion of capacity.

Carleson, however, in a different vein, showed [2] that for a class of kernels of the form

$$K(x) \sim 1/|x|(\log 1/|x|)^{1+\beta} \quad (\beta > 0), \tag{7.10}$$

the following inequality holds

$$C_K\{x \in [-\pi, \pi] : \theta_f(x) > \lambda\} \leq (c/\lambda^2)\|f\|_K^2, \tag{7.11}$$

where $\theta_f(x)$ denotes the Hardy-Littlewood function of f, that is, the function

$$\theta_f(x) = \sup\{\frac{1}{h}\int_{-h}^{h} |f(x+t)|dt: h > 0\}.$$

The inequality in (7.11) has several interesting consequences. Among other things, one can show from (7.11) that a function $f \in \mathcal{H}_K$, with K of the form (7.10), has a Fourier series whose Abel and Cesaro sums converge except on a set of K-capacity zero.

It is not difficult to show that the condition $f \in \mathcal{H}_K$, when K is of the form (7.10), is equivalent to one of the type (7.2) with (7.1) holding true. Thus Carleson's inequality seems to apply precisely when our Lemma 3.1 fails to apply.

It is clear then that it might be worthwhile to explore some further consequences of Carleson's inequality. To this end let us suppose that $f(x)$ is monotone decreasing in $[-\pi, \pi]$ and that $f \in \mathcal{H}_K$. In this case, for each $\lambda > 0$, the set

$$E_\lambda = \{x: f(x) > \lambda\}$$

is an interval, and the inequality

$$C_K(E_\lambda) \leqslant (c/\lambda^2)\|f\|_K^2 \tag{7.12}$$

gives an upper bound for the size of this interval. To see this, let us introduce for each kernel $K(x)$ the associate kernel

$$\overline{K}(x) = \frac{1}{|x|}\int_0^{|x|} K(t)dt \quad (x \in [-\pi, \pi]),$$

and let $\varphi(y)$ denote the inverse of the function

$$1/\overline{K}(x/2) \quad (\text{for } x > 0)\ ^*$$

If E is an interval and $|E|$ indicates its length, it is easily seen that the measure

$$d\mu = (1/|E|\overline{K}(|E|/2))\chi_E(x)dx,$$

where $\chi_E(x)$ denotes the indicator of E, is admissible for the supremum problem in (7.9). Thus we must have

$$1/\overline{K}(|E|/2) \leqslant C_K(E), \tag{7.13}$$

or better yet

$$|E| \leqslant \varphi(C_K(E)). \tag{7.14}$$

Combining (7.14) and (7.12) we deduce that for each $\lambda > 0$

$$m\{x: f(x) > \lambda\} \leqslant \varphi((c/\lambda^2)\|f\|_K^2). \tag{7.15}$$

In other words, one of the consequences of Carleson's inequality is an *a priori* rate of decay for the $m\{x: f(x) > \lambda\}$, which is the same for all functions $f \in \mathcal{H}_K$ with $\|f\|_K^2$ bounded above by the same constant, say B.

It is true that we have deduced this result rule for *monotone* functions $f(x)$, but *if* the conjecture of §5 is true, then (7.15) must hold for $f^*(x)$ and then must also hold for all $f \in \mathcal{H}_K$.

This is all to the good, but it might not be clear yet in what way the inequality in (7.15) is related to our inequality in (3.7), beyond the fact that both follow from similar conditions on f.

---

* If $K(x)$ decreases for $0 < x < \pi$, so does $\overline{K}(x)$ so $\varphi(y)$ is well defined.

§8. THE GENERAL BASIC LEMMA. To see the connection between (7.15) and (3.7) we must free K from being one of the Carleson kernels (7.10). As a matter of fact, K must be allowed to vary even beyond the domain usually adopted in the theory of Dirichlet spaces.

Indeed, (7.15) becomes essentially of the type (3.8) precisely when the kernel K is bounded. In the theory of Dirichlet spaces K is never taken to be bounded, for the corresponding capacity theory becomes rather dull then: the empty set is the only set of capacity zero.

Assuming that (7.11) holds for all such kernels, from (7.15) we easily obtain

$$m\{x: f(x) > \lambda\} = 0$$

as soon as

$$(c/\lambda^2)\|f\|_K^2 \leqslant 1/\overline{K}(0).$$

In other words

$$\text{(8.1)} \qquad \text{ess sup } f(x) \leqslant c\|f\|_K\sqrt{\overline{K}(0)}.$$

And we can easily deduce from this that

$$\text{ess sup } f(x) - \text{ess inf } f(x) \leqslant 2c\|f\|_K\sqrt{\overline{K}(0)}.$$

This inequality is of course only of the type (5.4), but a suitable change of scale argument would again yield an upper bound for the modulus of continuity of f.

These considerations might suggest that we should try extending Carleson's inequality (7.11) to all such kernels.

Interesting as this task may be, (7.11) is not really needed here, for it is easy to see that (8.1) is an almost immediate consequence of Schwarz's inequality.

What we want to do here is obtain our estimate in a form which will lead us to the desired extension of Lemma 3.1. To this end, let

$$K(x) \sim \Sigma\, e^{inx}/\lambda_n \in \mathcal{L}_1(-\pi, \pi)$$

and $\lambda_n = \lambda_{-n} \geqslant 0$, with $K(x)$ increasing as $|x|$ decreases in $(0, \pi)$. Set again

$$\overline{K}(x) = \frac{1}{x}\int_0^x |K(t)|dt.$$

Now if

$$\text{(8.2)} \qquad \|f\|_K = \sqrt{\sum_n |c_n|^2\lambda_n} < \infty.$$

set for a given $\lambda > 0$

$$E = \{x: f(x) \geqslant \lambda\}, \quad |E| = m(E).$$

By Parseval's relation we then obtain

$$\lambda\,|E| \leqslant \int_{-\pi}^{\pi} f(x)\chi_E(x)dx = 2\pi \sum_n c_n\gamma_n$$

where

$$\gamma_n = \frac{1}{2\pi}\int_{-\pi}^{\pi} e^{-inx}\chi_E(x)dx.$$

So, Schwarz's inequality yields

$$\lambda|E| \leqslant 2\pi\|f\|_K\sqrt{\sum_n |\gamma_n|^2/\lambda_n}. \tag{8.3}$$

Now note that since $K(x) \in \mathcal{L}_1$, the function

$$\int_{-\pi}^{\pi} K(x-y)\chi_E(y)dy$$

is square integrable as well, so again Parseval's relation gives

$$(2\pi)^2\, \Sigma\, |\gamma_n|^2/\lambda_n = \int_E \int_E K(x-y)dxdy. \tag{8.4}$$

Combining (8.3) and (8.4) with the simple inequality

$$\int_E \int_E K(x-y)dxdy \leqslant |E|^2 K(|E|/2)$$

we finally deduce

$$\lambda \leqslant \|f\|_K\sqrt{K(|E|/2)}. \tag{8.5}$$

We want to rewrite this inequality in a more suggestive form. To this end note that if $f^*(x)$ denotes the nondecreasing rearrangement of $f(x)$ in $[0, 2\pi]$ and we set $\lambda = f^*(x)$ then

$$m\{t: f(t) \geqslant \lambda\} = x.$$

Using this fact in (8.5) we obtain

$$f^*(x) \leqslant \|f\|_K\sqrt{K(x/2)} \text{ for every } 0 < x \leqslant 2\pi. \tag{8.6}$$

In other words for all functions $f \in \mathcal{L}_2(-\pi, \pi)$ for which $\|f\|_K^2 \leqslant B$ we have an *a priori* bound for the *rate of increase* of $f^*(x)$ as $x \downarrow 0$. If $K(0) < \infty$, then (8.6) gives an *a priori* bound for $f^*(x)$ and thus a bound like (8.1) for $f(x)$ itself.

This observation will lead us naturally to the appropriate generalization of Lemma 3.1.

We still need only a very small modification of (8.6). Note that a condition such as

$$B(f) = \int_{-\pi}^{\pi}\int_{-\pi}^{\pi} |(f(x) - f(y))/p(x-y)|^2 dxdy < \infty,$$

can be rewritten in the form (8.2) with $\lambda_0 = 0$. This implies that for functions having zero mean in $[-\pi, \pi]$ and a suitable $K(x)$ we have

$$\|f\|_K = \sqrt{B(f)}.$$

This given, from (8.6) we can easily derive the inequality

$$\text{(8.7)} \qquad f^*(x) - \frac{1}{2\pi}\int_{-\pi}^{\pi} f(t)dt \leq \sqrt{B(f)}\overline{K}(x/2).$$

It turns out that the arguments leading to Lemma 4.1 combined with our conjecture of §5 lead precisely to this type of inequality.

Let us go back to the proof of Lemma 4.1 with the same assumptions concerning f, Ψ and p. Here again $\{t_n\}$ is defined as in (4.3). Furthermore, for given $0 < x < y \leq 1/2$ let us also introduce the triangle

$$\Delta = \{(s,t): y \leq s \leq y + x,\ s - y \leq t \leq x\}.$$

Reasoning as we did for each Δ_n, we obtain

$$B/2 \geq \iint_{\Delta} \Psi((f(s) - f(t))/p(s-t))dsdt \geq (1/2)x^2\Psi((f(x) - f(y))/p(y)),$$

and, inverting Ψ

$$\text{(8.8)} \qquad f(x) - f(y) \leq \Psi^{-1}(B/x^2)p(y).$$

For given $0 < x \leq 1/2$ pick N so that

$$t_{N+1} < x \leq t_N.$$

Trivially

$$f(x) - f(1/2) = f(x) - f(t_{N-1}) + \sum_{\nu=0}^{N-2} [f(t_{\nu+1}) - f(t_\nu)],$$

Thus by (4.7) and (8.8) with $y = t_{N-1}$

$$f(x) - f(1/2) \leq \Psi^{-1}(B/x^2)p(t_{N-1}) + 4\sum_{\nu=0}^{N-2}\int_{t_{\nu+2}}^{t_{\nu+1}} \Psi^{-1}(B/u^2)dp(u).$$

However, since

$$p(t_{N-1}) = 2p(t_N) = 4p(t_{N+1}) \leq 4p(x)$$

we finally get

$$\text{(8.9)} \qquad f(x) - f(1/2) \leq 4\Psi^{-1}(B/u^2)p(x) + 4\int_x^{1/2} \Psi^{-1}(B/u^2)dp(u).$$

Notice now that since

$$\Psi^{-1}(B/u^2)p(x) \leq \int_0^x \Psi^{-1}(B/u^2)dp(u),$$

(8.9) gives still (4.8) when

$$\int_0^1 \Psi^{-1}(B/u^2)dp(u) < \infty. \tag{8.10}$$

Otherwise, (8.9) yields an *a priori* bound for growth of $f(x)$ as $x \downarrow 0$. A bound similar to (8.9) can be obtained for $1/2 < x < 1$ by working with $-f(1 - x)$ in place of $f(x)$.

To sum up, we can conclude that if the conjecture of §5 holds for a pair of functions $\Psi(u)$ and $p(u)$ then the following result holds as well:

LEMMA 8.1. Let $f(x)$ be the measurable on $[0, 1]$ and suppose that

$$\int_0^1 \int_0^1 \Psi((f(s) - f(t))/p(s - t))dsdt \leq B < \infty \tag{8.11}$$

then for $0 < x \leq 1/2$ we have

$$f^*(x) - f^*(1/2) \leq 4\Psi^{-1}(B/x^2)p(x) + 4 \int_x^{1/2} \Psi^{-1}(B/u^2)dp(u) \tag{8.12}$$

while for $1/2 \leq x \leq 1$ we have

$$f^*(1/2) - f^*(x) \leq 4\Psi^{-1}(B/(1 - x^2))p(1 - x) + + 4 \int_{1-x}^{1/2} \Psi^{-1}(B/u^2)dp(u). \tag{8.13}$$

Remarks. It is interesting to see what this lemma yeilds in case (8.11) is one of the classical conditions.

For instance suppose f satisfies Dirichlet's condition

$$\int_0^1 \int_0^1 ((f(x) - f(y))/(x - y))^2 dxdy \leq B < \infty. \tag{8.14}$$

Then from (8.12) we get

$$f^*(x) - f^*(1/2) \leq 4\sqrt{B} + 4\sqrt{B} \log 1/2x.$$

So that for every $0 < \gamma < 1$ we easily obtain

$$\int_0^{1/2} \exp \gamma((f^*(x) - f^*(1/2))/4\sqrt{B})dx \leq e^\gamma/2(1 - \gamma).$$

If we combine this inequality with the one we obtain from (8.13) and make use of property (5.2) we can finally deduce that

$$\int_0^1 \exp \gamma |(f(x) - f^*(1/2))/4\sqrt{B}| dx \leq e^\gamma/(1 - \gamma). \tag{8.15}$$

In other words, any function $f \in \mathcal{L}_2$ *which satisfies (8.14) must necessarily be in all $\mathcal{L}_2$ classes* $(p < \infty)$. It turns out that this is actually true! Indeed, if we work with the equivalent condition

$$\sum_n |c_n|^2 n < \infty,$$

the fact that $f \in \mathcal{L}_p$ (for every $p < \infty$) is a simple consequence of the Hausdorff-Young inequality.

Similarly, from the condition

$$\int_0^1 \int_0^1 ((f(x) - f(y))/|x - y|^a)^2 \, dxdy \leqslant B$$

for $0 < a < 1$, we obtain the estimate

$$\int_0^1 [(f(x) - f^*(1/2))/4\sqrt{B}]^p dx \leqslant 2^{p(1-a)}/(1 - a)^p [1 - p(1 - a)]$$

Thus $f \in \mathcal{L}_p$ for every $p < 1/(1 - a)$. This also follows from the Hausdorff-Young inequality.

A study of the implications of Lemma 8.1 in those cases that are manageable by other methods reveals that (8.12) (and (8.13)) become quite sharp when $\Psi(u) \uparrow \infty$ very rapidly as $u \uparrow \infty$.

For $\Psi(u) = u^2$ however, and $p(u) \sim |u| (\log 1/|u|)^a$ the estimates given by (8.12) are not as good as those that can be obtained from (8.7). This should not be very surprising, for in those cases, the inequality

$$f(t_{n+1}) - f(t_n) \leqslant 4 \int_{t_{n+2}}^{t_{n+1}} \Psi^{-1}(B/u^2) dp(u),$$

which was instrumental in proving (8.9), becomes very crude.

It is possible to obtain sharper inequalities than (8.12) and (8.13). However, since the arguments needed for this purpose become rather technical, we shall have to refer the reader to [8] for further details.

The main point we wanted to make here is that it is worth investigating further whether or not our conjecture is true.

§9. RODEMICH'S APPROACH TO THE CONJECTURE AND TAYLOR'S GRAPH THEORETICAL RESULT. The formulation of Lemma 8.1 and the discovery that some of the consequences of this lemma are indeed true brought new stimulus to our efforts at establishing the conjecture.

Thus at the beginning of the summer of 1969, E. Rodemich was led to a combinatorial problem whose solution by H. Taylor yielded a proof of the inequality (5.3) for all p satisfying (3.5) and all Ψ which in addition to (3.4) have the property that the function $\Psi(e^u)$ is convex.

We shall give here only a description of this work in its basic lines and refer the reader to the original papers [8] and [15] for further details.

The point of departure is the following simple but powerful observation of E. Rodemich. If $\Psi(u)$ has a continuous derivative, for any Δ and any u we have

$$\Psi(\Delta/p(u)) = \Psi(\Delta/p(1)) + \int_u^1 \Psi'(\Delta/p(\delta))(\Delta/(p(\delta))^2)dp(\delta)$$

Thus if we set

$$\Phi(u) = \Psi'(u)u \tag{9.1}$$

we can write

$$\Psi(\Delta/p(u)) = \Psi(\Delta/p(1) + \int_u^1 \Phi(\Delta/p(\delta))\frac{dp(\delta)}{p(\delta)}.$$

Substituting this relation in the integral

$$I_{\Psi, p}(f) = \int_0^1 \int_0^1 \Psi((f(x) - f(y))/p(x - y))dxdy,$$

an application of Fubini's theorem yields

$$\begin{aligned} I_{\Psi, p}(f) &= \int_0^1 \int_0^1 \Psi((f(x) - f(y))/p(1))dxdy + \\ &+ \int_0^1 [\textstyle\iint_{|x-y|\leq\delta} \Phi((f(x) - f(y))/p(\delta))dxdy]\frac{dp(\delta)}{p(\delta)}. \end{aligned} \tag{9.2}$$

For convenience, given a function $\Phi(u)$ defined, nonnegative and symmetric in $(-\infty, +\infty)$, let us introduce the functionals

$$J_{\Phi, \delta}(f) = \iint_{|x-y|\leq\delta} \Phi(f(x) - f(y))dxdy \quad (0 < \delta < 1), \tag{9.3}$$

where the integration is to be carried out only inside the square $S = [0, 1] \times [0, 1]$.

This given, it is easy to see from (9.2) that the inequality

$$I_{\Psi, p}(f^*) \leq I_{\Psi, p}(f)$$

must necessarily follow for all $\Psi(u)$ such that

$$u\Psi'(u) \tag{9.4}$$

is monotone increasing for $u > 0$, as soon as we show that

$$J_{\Phi, \delta}(f^*) \leq J_{\Phi, \delta}(f) \tag{9.5}$$

holds for all $0 < \delta < 1$ and all $\Phi(u)$ which are nonnegative symmetric and monotone increasing for $u \geq 0$. To simplify our exposition let the class of all $\Phi(u)$ satisfying the latter conditions be denoted by $\mathfrak{M}$.

In studying the integrals (9.3) one is naturally led to consider the analogous discrete expressions

$$(9.6) \qquad Q_{\Phi, M}(f) = \sum_{|i-j| \leq M} \Phi(f_i - f_j),$$

where $f = (f_1, f_2, \ldots, f_N)$ is a vector of real constants, $1 \leq M \leq N$, and the sum is of course carried out only over those couples (i, j) in the square $\mathfrak{S}$ of lattice points

$$\mathfrak{S} = \{(i, j): i = 1, 2, \ldots, N;\ j = 1, 2, \ldots, N\}.$$

In analogy with the continuous case let, for a given vector $f = (f_1, f_2, \ldots, f_N)$,

$$f^* = (f_1^*, f_2^*, \ldots, f_N^*)$$

denote the vector obtained by rearranging the components of f in decreasing order.

At this point, it is not difficult to see that if the discrete version of (9.5), namely

$$(9.7) \qquad Q_{\Phi, M}(f^*) \leq Q_{\Phi, M}(f)$$

holds for all f, all $1 \leq M \leq N$ and all $\Phi \in \mathfrak{M}$ then, by a passage to the limit, (9.5) must hold as well.

It turns out that (9.7) leads to a very beautiful combinatorial problem. To see this we shall have to rewrite (9.7) in a slightly different form.

First of all, let $\overline{\mathfrak{S}}$ denote the upper diagonal part of $\mathfrak{S}$ namely

$$\overline{\mathfrak{S}} = \{(i, j): 1 \leq i \leq j \leq N\}.$$

Further, if we set $a = (i\,j)$ (no comma), we mean the point (i, j) or (j, i) according as $i \leq j$ or $j < i$. With this convention, let for given $1 \leq M \leq N$

$$\mathscr{A}_M = \{a = (i\,j): 0 \leq j - i \leq M\}.$$

Now, for a fixed vector of real constants $(y_1, y_2, \ldots, y_N)$ such that

$$y_1 \geq y_2 \geq \ldots \geq y_N$$

and a fixed $\Phi \in \mathfrak{M}$ set for every $a \in \overline{\mathfrak{S}}$

$$(9.8) \qquad \varphi(a) = \Phi(y_i - y_j).$$

Moreover, given a permutation $\sigma = (\sigma_1, \sigma_2, \ldots, \sigma_N)$ of $(1, 2, \ldots, N)$ let, for $a = (i\,j)$

$$(9.9) \qquad \sigma a = (\sigma_i\, \sigma_j).$$

This formula defines a one-to-one transformation of $\overline{\mathfrak{S}}$ into itself, and using the customary notation set

$$\sigma\mathscr{A}_M = \{b: b = \sigma a,\ a \in \mathscr{A}_M\}.$$

All this being given, it is not difficult to see that (9.7) will necessarily follow if we show that the inequality

$$\sum_{a\in\mathcal{S}_M} \varphi(a) \leqslant \sum_{a\in\sigma\mathcal{S}_M} \varphi(a)$$

holds for all $y_1 \geqslant y_2 \geqslant \ldots \geqslant y_N$, all $\Phi \in \mathfrak{M}$ and all permutations $\sigma = (\sigma_1, \ldots, \sigma_N)$ of $(1, 2, \ldots, N)$.

Observe now that, for any given $a = (i\,j)$ (with $i \leqslant j$) we must have

$$\varphi(a) \leqslant \varphi(a')$$

for all $a' = (i'\,j')$ in the "rectangle"

$$R_a = \{a' \colon 1 \leqslant i' \leqslant i;\ j \leqslant j' \leqslant N\}. \tag{9.11}$$

This fact follows easily from (9.8) and our monotonicity assumptions on Φ and y.

This given, a tempting way to establish (9.10) would be to pair off each point $a \in \mathcal{S}_M$ with a distinct point $a' \in \sigma\mathcal{S}_M$, and make sure that every time we also have

$$a' \in R_a.$$

This idea led E. Rodemich to a further version of the conjecture. Let us introduce a partial ordering of $\overline{\mathfrak{S}}$ by setting

$$a \prec \pi(a) \text{ for every } a \in \mathcal{S}_M. \tag{9.12}$$

CONJECTURE 9.1. Given any $1 \leqslant M \leqslant N$ and any permutation σ we can always find a one-to-one map $\pi(a)\colon \mathcal{S}_M \to \sigma\mathcal{S}_M$ such that

$$a \prec \pi(a) \text{ for every } a \in \mathcal{S}_M.$$

Clearly, if this conjecture is true, we can then write

$$\sum_{a\in\sigma\mathcal{S}_M} \varphi(a) = \sum_{a\in\mathcal{S}_M} \varphi(\pi(a))$$

But since $\pi(a) \succ a$ we necessarily have

$$\varphi(\pi(a)) \geqslant \varphi(a) \text{ for every } a \in \mathcal{S}_M$$

and (9.10) immediately follows.

The remarkable fact is that the implication also goes the other way around. In other words, if (9.10) holds for all such φ's then the conjecture must also hold true. This however is not as easy to show and since it is not needed it will be omitted.

To establish the conjecture we must be able to select in each rectangle R_a with $a \in \mathcal{S}_M$ a "*representative*" point $a' \in \sigma\mathcal{S}_M$ so that *distinct* rectangles have *distinct representatives* in $\sigma\mathcal{S}_M$.

However, in combinatorial analysis there is a classical theorem due to Phillip Hall which covers precisely this kind of situation and gives necessary and sufficient conditions for such a selection to be possible.

For the benefit of the reader who may not be acquainted with this result we shall state it in the specific form needed here. Let X be a finite set of points. Let $\mathfrak{F}$ be a family of subsets F of X each intersecting a certain fixed subset Y of X. We say that we have a *"system of distinct representatives for $\mathfrak{F}$ in S"* if we have a one-to-one map π of $\mathfrak{F}$ into X such that for each $F \in \mathfrak{F}$

$$\pi(F) \in F \cap Y.$$

THEOREM 9.1 (P. Hall). Given X, $\mathfrak{F}$ and Y it is possible to select a system of distinct representatives for $\mathfrak{F}$ in Y if and only if for all choices $F_1, F_2, \ldots, F_k \in \in \mathfrak{F}$ we have

$$|(F_1 \cup F_2 \cup \ldots \cup F_k) \cap Y| \geq k \; * \tag{9.13}$$

A very readable proof of this theorem may be found in [13].

Phillip Hall's theorem immediately yields another version of our conjecture. We need only translate (9.13), when we take $X = \overline{\mathcal{S}}$,

$$\mathfrak{F} = \{R_a : a \in \mathcal{A}_M\}$$

and

$$Y = \sigma\mathcal{A}_M.$$

However, (9.13) must be used in a judicious way. Indeed, if $a_1, a_2, \ldots, a_\ell$ are points in $\mathcal{A}_M$, the union

$$U = \bigcup_{i=1}^{\ell} R_{a_i}$$

can also be written as a union of as many as

$$|U \cap \mathcal{A}_M|$$

distinct rectangles R_a with $a \in \mathcal{A}_M$.

Thus the new conjecture reads:

CONJECTURE 9.2. For any $x \leq M \leq N$, any permutation σ and any union $U = \bigcup_{i=1}^{\ell} R_{a_i}$ with $a_1, a_2, \ldots, a_\ell \in \mathcal{A}_M$ we have

$$|U \cap \sigma\mathcal{A}_M| \geq |U \cap \mathcal{A}_M|. \tag{9.14}$$

Since the rectangles R_a sort of "spread away" from the diagonal of $\mathcal{S}$ and $\sigma\mathcal{A}_M$

* If A is a finite set, by $|A|$ we denote the number of points in A.

is most concentrated about the diagonal precisely when σ is the identity permutation, the inequality (9.14) looks quite plausible.

In the form (9.2) the conjecture was presented to H. Taylor by E. Rodemich at the beginning of the summer of 1969. A very short time later H. Taylor discovered a remarkably ingenious proof of it using a graph theoretical approach. This proof will be found in [15].

As a result of Taylor's work, the conjecture of §5 must then hold true whenever $\Psi(e^u)$ is convex, and so for such Ψ's Lemma 8.1 must also hold. In addition we get also a new proof of Lemma 3.1, with some improvement on the bounds at least for this special class of Ψ's.

It is to be noted that since the power functions $\Psi(u) = |u|^p$ belong to this class the classical Dirichlet condition

$$\int_0^1 \int_0^1 ((f(x) - f(y))/(x - y))^2 \, dxdy < \infty$$

as well as any of the conditions

$$\int_0^1 \int_0^1 ((f(x) - f(y))/|x - y|^\alpha)^p \, dxdy < \infty$$

are all within the range of applications of the theory presented here.

§10. PRESTON'S WORK. In a brilliant thesis [12] C. Preston was able to extend Carleson's inequality

$$C_K\{x: \theta_f(x) > \lambda\} \leq (c/\lambda^2)\|f\|_K^2 \tag{10.1}$$

to the class of kernels

$$K(x) \sim \Sigma \, e^{inx}/\lambda_n$$

satisfying the following conditions

(10.2)

(i) $K(x)$ is continuous for $0 < |x| \leq \pi$,

(ii) $K(x) \geq 0$,

(iii) $K(x) = K(-x)$,

(iv) $K(s) \geq K(t)$ for $0 < s \leq t \leq \pi$,

(v) K is convex in $(0, \pi]$,

(vi) K is in $\mathcal{L}_1[-\pi, \pi]$

(vii) $K(x) = 0(K(2x))$ as $|x| \to 0$

For convenience let us denote here by $\mathcal{PC}$ the class of K's satisfying in addition to (i) - (vii) in (10.2) also the condition

(10.3) (viii) The sequence $\{1/(\lambda_n + \lambda)\}$ is positive definite for every $\lambda \geq 0$.

The exact form of conditions (i) - (v) is not essential in Preston's work, however, these conditions make K(x) good to work with.

Conditions (vii) and (viii) play a special role. Indeed, (vii) assures that the corresponding "potential" theory admits a "Harnack inequality", while condition (viii) is necessary and sufficient so that the norm

$$\|f\|_K = (\Sigma\, |c_n|^2 \lambda_n)^{1/2}$$

will be equivalent to one of the form

$$B(f) = [\int_{-\pi}^{\pi} |f|^2 dx + \int_{-\pi}^{\pi}\int_{-\pi}^{\pi} [(f(x) - f(y))/p(x - y)]^2 dxdy]^{1/2}.$$

One very important fact shown by Preston is that given any function $f \in \mathcal{L}_2[-\pi, \pi]$ there is a K satisfying conditions (i) - (viii) such that

$$\|f\|_K = (\Sigma\, |c_n|^2 \lambda_n)^{1/2} < \infty.$$

In other words

$$\bigcup_{K\in\mathcal{PC}} \mathcal{H}_K = \mathcal{L}_2[-\pi, \pi].$$

This result and a self-contained development of this theory may be found in [12].

A very short derivation of (10.1) may be found in Preston's article in these proceedings. Preston's methods are quite different from Carleson's -- they are very natural and elegant. This fact may be obscured by Preston's concise style of writing.

For these reasons in this section we shall make a few remarks which we hope will be of help in understanding Preston's line of reasoning.

First of all, to put (10.1) in the proper prospective, let us assume that for a fixed $K \in \mathcal{PC}$ and for some given (perhaps bounded) Borel function f(x) we have

$$\Sigma\, |c_n|^2 \lambda_n < \infty$$

where

$$c_n = \frac{1}{2\pi} \int_{-\pi}^{\pi} f(x) e^{-inx} dx.$$

If $\mu(x)$ is a Borel measure, and

$$\gamma_n = \frac{1}{2\pi} \int_{-\pi}^{\pi} e^{inx} d\mu(x)$$

then it follows that if

$$(10.3) \qquad I_K(\mu,\mu) = \int_{-\pi}^{\pi}\int_{-\pi}^{\pi} K(x-y)d\mu(x)d\mu(y) < \infty$$

the series

$$(10.4) \qquad S = 2\pi \sum_n c_n\gamma_n$$

is absolutely convergent.

This is a consequence of Schwarz's inequality and the easily proved (but not trivial) fact that for our kernels we have

$$I_K(\mu,\mu) = (2\pi)^2 \sum |\gamma_\nu|^2/\lambda_\nu.$$

This given one might carelessly assert that

$$(10.5) \qquad S = \int_{-\pi}^{\pi} f(x)d\mu(x).$$

However, chances are, this is false.

Indeed, although the right-hand side of (10.5) is well defined for any bounded Borel function $f(x)$, it may not be equal to S for the simple reason that even if we change f on a set of measure zero (which could very well be the support of μ) S does not change.

Curiously enough, however, if following Zygmund's notation we set

$$\sigma_n(x,f) = \sum_{|\nu|\leqslant n} (1 - |\nu|/(n+1))c_\nu e^{i\nu x},$$

then we clearly have

$$(10.6) \qquad \lim_{n\to\infty} \int_{-\pi}^{\pi} \sigma_n(x,f)d\mu = 2\pi \sum c_\nu\sigma_\nu.$$

This fact should suggest that within the equivalence class (in $\mathcal{L}_2$) determined by f there is a representative $\tilde{f}$ for which

$$(10.7) \qquad \int_{-\pi}^{\pi} \tilde{f}(x)d\mu(x) = S$$

holds true, and that $\sigma_n(x,f)$ converges to that representative in such a fashion that we may derive (10.7) from (10.6).

We see that we have here a situation analogous to the one we have encountered for the Karhunen-Loève expansion. The Fourier series

$$\sum c_n e^{inx}$$

must be convergent in the Cesaro sense more often than one suspects, and to the ***right*** limit as well!

Some of the puzzle here can be removed by use of (10.1).

To see this, we point out that if for a Borel set E we have

$$C_K(E) = 0$$

then we must have

$$\mu(E) = 0 \text{ for every } \mu \text{ such that } I_K(\mu, \mu) < \infty.$$

This fact is a consequence of the classical identity:

$$C_K(E) = 1/\inf\{I_K(\mu, \mu): \mu(E) \geq 1\}. * \tag{10.10}$$

Using (10.1) it is easy to show that the Cesaro sums $\sigma_n(x, f)$ fail to converge only on a set of K-capactiy zero in x. Thus they converge a.e. $d\mu$ for any measure μ such that $I_K(\mu, \mu) < \infty$.

This given, by setting

$$\tilde{f}(x) = \begin{cases} \lim_{n\to\infty} \sigma_n(x, f) & \text{when it exists} \\ 0 & \text{otherwise,} \end{cases}$$

we can easily establish (10.7) from (10.6) at least when f is bounded. However, once this is done, the extension of (10.7) to all $f \in \mathcal{H}_K$ is easily carried out. **

This is fine but a suspicion should remain that some further underlying facts are to be discovered here concerning the convergence behavior of the sequence $\sigma_n(x, f)$ for $f \in \mathcal{H}_K$.

Let us now go over the proof of (10.1).

It is clear that we need only work with $\lambda = 1$. Preston works with the function

$$\theta_f(x) = \sup_h \frac{1}{h} \int_0^h f(x + t)dt$$

and starts with the classical observation that we can write

$$\{x: \theta_f(x) > 1\} = \sum_{\nu=1}^{\infty} I_\nu$$

where the I_ν's are disjoint intervals such that

$$\frac{1}{|I_\nu|} \int_{I_\nu} f(x)dx = 1. \tag{10.11}$$

---

* For a proof see [12].

** We should perhaps point out that for this last step we need (viii).

He then sets himself the task, given a fixed n, to find the function $f \in \mathcal{H}_K$ satisfying (10.11) for $\nu = 1, 2, \ldots, n$ and whose norm $\|f\|_K$ is the smallest. The hope here is that the norm of the extremal function might be related somehow to the capacity of the set

$$E = \sum_{\nu=1}^{n} I_\nu.$$

Preston then discovers that upon replacing (10.11) by the less restricting constraint

$$\frac{1}{|I_\nu|} \int_{I_\nu} f(x)dx \geqslant 1 \tag{10.12}$$

The resulting extremal problem becomes quite natural and indeed has a rather interesting solution. To present these facts we need to introduce some notation. First of all we shall set for

$$f \sim \Sigma\, c_n e^{inx}, \quad g \sim \Sigma\, d_n e^{inx},$$

$f, g \in \mathcal{H}_K$

$$(f, g)_K = \sum_\nu c_\nu \overline{d_\nu} \lambda_\nu. \tag{10.13}$$

We also set

$$(f, g) = 2\pi \sum_\nu c_\nu \overline{d_\nu} = \int_{-\pi}^{\pi} f\overline{g}dx.$$

And we observe that if

$$g = K * h = \int_{-\pi}^{\pi} K(x - t)h(t)dt$$

then

$$(f, g)_K = (f, h).$$

Further, we introduce the indicator functions

$$\chi_\nu(x) = \begin{cases} 1, & x \in I_\nu \\ 0, & x \notin I_\nu \quad (\nu = 1, 2, \ldots, n), \end{cases}$$

and their transforms

$$\theta_\nu(x) = K * \chi_\nu = \int_{I_\nu} K(x - t)dt \quad (\nu = 1, 2, \ldots, n).$$

We shall also need the basis dual to

$$\langle\theta_1, \theta_2, \ldots, \theta_n\rangle$$

with respect to the inner product (10.13). In other words the basis

$$\langle\theta_1^*, \theta_2^*, \ldots, \theta_n^*\rangle = \langle\theta_1, \theta_2, \ldots, \theta_n\rangle A$$

where A is the inverse of the matrix

$$B = \|(\theta_i, \theta_j)_K\| = \|(\chi_i, K * \chi_j)\| \quad *$$

This given, we can construct the projection $P(f)$ of a given $f \in \mathcal{H}_K$ into the linear span $L(\theta_1, \theta_2, \ldots, \theta_n)$ of $\theta_1, \theta_2, \ldots, \theta_n$ by the formulas

$$P(f) = \sum_{\nu=1}^{n} (\theta_\nu^*, f)_K \theta_\nu = \sum_{\nu=1}^{n} (\theta_\nu, f)_K \theta_\nu^*.$$

Note then that

$$\int_{I_\nu} P(f)dx = (P(f), \theta_\nu)_K = (f, \theta_\nu)_K = \int_{I_\nu} fdx$$

Thus if f satisfies (10.12), so does P(f). Pithagoras theorem however gives

$$\|f\|_K^2 = \|P(f)\|_K^2 + \|f - P(f)\|_K^2.$$

This shows that the solution of the extremal problem must lie in $L(\theta_1, \theta_2, \ldots, \theta_n)$. For f of the form

$$f = \sum_{\nu=1}^{n} c_\nu \theta_\nu = \sum_{\nu=1}^{n} d_\nu \theta_\nu^* \quad (d = Bc),$$

we have

$$(f, f)_K = \sum_{\nu,\mu=1}^{n} c_\nu c_\mu B_{\nu,\mu} = c^T B_c$$

and

$$\int_{I_\nu} fdx = d_\nu = \sum_{\mu=1}^{n} B_{\nu,\mu} c_\mu.$$

Thus we are reduced to the finite dimensional problem

$$\text{(10.14)} \qquad \inf_{\{Bc \geqslant m\}} c^T Bc$$

where we have set

---

* B is nonsingular since $I_1, I_2, \ldots, I_n$ are supposed to be disjoint and K is a positive definite kernel.

$$m = \begin{Bmatrix} |I_1| \\ |I_2| \\ \vdots \\ |I_n| \end{Bmatrix}.$$

Since B is strictly positive definite $|c^T Bc| \to \infty$ as $|c| \to \infty$ so the minimum clearly exists. Let us denote by $\mathring{c}$ the solution of (10.14). And let us set

$$\mathring{f} = \sum_{\nu=1}^{n} \mathring{c}_\nu \theta_\nu.$$

Preston establishes the following important properties of $\mathring{c}$:

(a) $\mathring{c}_\nu \geqslant 0,\ \nu = 1, 2, \ldots, n;$

(b) $\mathring{d}_\nu = \int_{I_\nu} \mathring{f} dx = \sum_{\mu=1}^{n} B_{\nu,\mu} \mathring{c}_\mu > m_\nu$ only if $\mathring{c}_\nu = 0.$ *

From (b) one easily derives

$$(\mathring{f}, \mathring{f})_K = \sum_{\nu=1}^{n} \mathring{c}_\nu \mathring{d}_\nu = \sum_{\nu=1}^{n} \mathring{c}_\nu m_\nu = \int_{-\pi}^{\pi} (\sum_{\nu=1}^{n} \mathring{c}_\nu \chi_\nu) dx.$$

In other words our extremal function $\mathring{f}$ is a "potential"

$$\mathring{f}(x) = \int_{-\pi}^{\pi} K(x-y) d\mu(x)$$

with respect to a measure

$$d\mathring{\mu}(x) = (\sum_{\nu=1}^{n} \mathring{c}_\nu \chi_\nu) dx$$

totally concentrated on $E = \sum_{\nu=1}^{n} I_\nu$, and furthermore we have

$$(\mathring{f}, \mathring{f})_K = \mathring{\mu}(E).$$

At this point it is natural to ask if the extremal measure thus obtained is admissible for one of the extremal problems associated with the calculation of the capacity of E.

Preston shows that this is the case.

Indeed, a further way to calculate the capacity of a compact set E is via the formula

$$C_K(E) = \inf \mu(E) \tag{10.15}$$

---

* For a proof see Preston's article in these proceedings.

where the "inf" is taken over those measures μ such that

$$\int_{-\pi}^{\pi} K(x-t)d\mu(t) \geqslant 1 \text{ in } E \text{ (except on a set of K-capacity zero).}$$

Taking this into account, from (10.15) Preston derives the inequality

$$C_K(E) \leqslant (1/C)\mathring{\mu}(E) = (1/C)(\mathring{f}, \mathring{f})_K.$$

by showing that for a universal C we have

$$(10.16) \qquad \mathring{f}(x) = \int_{-\pi}^{\pi} K(x-t)d\mathring{\mu}(t) \geqslant C \text{ for every } x \in E.$$

His idea in proving this fact goes as follows. We do know that

$$(10.17) \qquad \frac{1}{|I_\nu|}\int_{I_\nu} \mathring{f}(x)dx \geqslant 1 \text{ for } \nu = 1, 2, \ldots, n.$$

now $\mathring{f}$ is nonnegative and is a potential, thus if the intervals I_ν are sufficiently far apart from each other, a Harnack type inequality will be valid and will provide us with an estimate of the form

$$(10.18) \quad \max\{\mathring{f}(x): x \in I_\nu\} \leqslant C \min\{\mathring{f}(x): x \in I_\nu\}, \; \nu = 1, 2, \ldots, n$$

with a universal C.

It is easy to see how (10.18) and (10.17) combined do yield (10.16).

This is Preston's proof of (10.1) in its essential lines. To complete the arguments we still have to put ourselves into a position where (10.18) does hold. This requires showing that some of the intervals I_ν can be omitted without too much loss in capacity. Further, an additional lemma is needed which controls the change in the capacity of E if each of its intervals is stretched by the same factor. These considerations and the further details omitted here may be found in the aforementioned works of C. Preston.

§11. SOME FINAL COMMENTS. Before closing we would like to mention some directions in which the work presented here might be extended.

First of all, it would be interesting to see what implication we can draw from Lemma 8.1 concerning the behavior of the Karhunen-Loève expansion when the condition

$$(11.1) \qquad \int_0^1 \sqrt{\log 1/n}\, dp(u) < \infty$$

of Theorem 6.1 does not hold.

We know in this case that the paths of the process $X_t(\omega)$ need not be continuous. However, we do believe that the partial sums

$$X_t^{(n)}(\omega) = \sum_{\nu=1}^{n} \sqrt{\lambda_\nu}\theta_\nu \varphi_\nu(t)$$

should exhibit some regularity. Indeed one might venture to say that they a.s. will fail to converge only on some very small set -- smaller in some sense than a set of measure zero, perhaps only on a set of capacity zero with respect to a suitable kernel.

One should also be able to obtain from Lemma 8.1 some kind of estimate for the remainder

$$X_t(\omega) - X_t^{(n)}(\omega) = \sum_{\nu=n+1}^{\infty} \sqrt{\lambda_\nu}\theta_\nu\varphi_\nu(t),$$

even when (11.1) fails to hold.

Another direction of further work could be to try and extend Lemma 3.1 as well as Lemma 8.1 to higher dimensions. By Fourier analytical techniques it is easy to show that in the cases

$$\Psi(u) = u^2, \ p(u) = |u|^{a}$$

such an extension can be carried out.

In this connection it might be worth mentioning here the class of functions of "bounded mean oscillation" introduced by F. John and L. Nirenberg [10] and extended by N. G. Meyers [9].

According to John and Nirenberg a function $f(x)$, defined on the n-dimensional and I_0, is said to have "bounded mean oscillation" if and only if for each subcube $I \subset I_0$ there is a constant m_I such that

$$\sup\{\tfrac{1}{|I|}\textstyle\int_I |f(x) - m_I| dx_1 dx_2 \ldots dx_n : I \subset I_0\} < \infty. \tag{11.2}$$

Actually there is an equivalent but more elegant way to state this condition, namely

$$\sup\{\tfrac{1}{|I|^2}\textstyle\int_I\int_I |f(x) - f(y)|\, dxdy: I \subset I_0\} < \infty. \tag{11.3}$$

In this form (11.2) appears to be a natural extension of the Diriclhet, one-dimensional, classical condition

$$\int_0^1\int_0^1 |(f(x) - f(y))/(x - y)|^2 dxdy < \infty,$$

for it is easy to see that for subcubes $I \subset I_0 = [0, 1]$

$$\frac{1}{|I|^2}\int_I\int_I |f(x) - f(y)|\, dxdy \leqslant \sqrt{\int_{I_0}\int_{I_0} |(f(x) - f(y))/(x - y)|^2 dxdy}$$

Similar conclusions can be drawn by writing N. G. Meyer's [9] condition in the form

$$\sup\{|I|^{-(2+a)}\int_I\int_I |f(x) - f(y)|\, dxdy: I \subset I_0\} < \infty.$$

We should point out that the methods of §8 can be used to derive a very simple proof of the John-Nirenberg and Meyers results for the one-dimensional case.

This given, the suspicion should arise that there might very well be a wider theory, free of dimension restrictions which include as special cases the results presented here as well as those of John-Nirenberg and Meyers.

REFERENCES

1. L. Carleson, Selected Problems on Exceptional Sets, Van Nostrand, 1967.
2. L. Carleson, Maximal functions and capacities, Ann. Inst. Fourier, 15 (1965), pp. 59-64.
3. J. Delporte, Fonctions aléatoires presque surement continues sur un intervalle fermé, Ann. Inst. Henry Poincaré, B 1 (1964-65), pp. 111-215.
4. J. L. Doob, Stochastic Processes, John Wiley and Sons, 1960.
5. X. Fernique, Continuité des processus Gaussiens, C. R. 258 (1964), pp. 6058-6060.
6. O. Frostman, Potentiel d'équilibre et capacité des ensembles avec quelques applications à la théorie des fonctions, Thesis, Lunds Univ. Mat. Sem., 3 (1935) pp. 1-118.
7. A. M. Garsia, E. Rodemich and H. Rumsey, Jr., A real variable lemma and the continuity of paths of some Gaussian processes, Journal of Math. and Mech., to appear.
8. A. M. Garsia and E. Rodemich, A result in the theory of graphs and the smoothness of functions satisfying certain integral inequalities, to appear.
9. N. G. Meyers, Mean oscillation over cubes and Hölder continuity, Proc. Amer. Math. Soc., 15 (1964), pp. 717-721.
10. F. John and L. Nirenberg, On functions of bounded mean oscillation, Comm. Pure and Appl. Math., 14 (1961), pp. 415-426.
11. R. E. A. C. Paley and N. Wiener, Fourier transforms in the complex domain, Amer. Math. Soc. Colloquium Publications, XIX (1934).
12. C. J. Preston, A theory of capacities and its application to some convergence results, Advances in Mathematics, to appear.
13. H. J. Reyser, Combinatorial mathematics, Carus Monograph, 14 (1963).

    E. M. Stein, Singular integrals, harmonic functions and differentiability properties of functions of several variables, Singular Integrals, Proc. of Symposia in Pure Mathematics, 10 (1967), pp. 316-335.
15. H. Taylor, Rearrangements of incidence tables, J. of Math. Anal. and Applications, to appear.
16. A. Zygmund, Trigonometric Series, Cambridge U. Press, 1959.